Longcase Painted Dials

Longcase Painted Dials

Their History and Restoration

M.F. TENNANT

Photography by D.A. Tennant

N.A.G. Press
an imprint of Robert Hale · London

First published in Great Britain 1995

ISBN 0 7198 0260 1

Robert Hale Limited
Clerkenwell House
Clerkenwell Green
London EC1R 0HT

2 4 6 8 10 9 7 5 3 1

Photoset in Ehrhardt by
Derek Doyle & Associates, Mold, Clwyd.
Printed in Great Britain by
St Edmundsbury Press Ltd, Bury St Edmunds, Suffolk.
Bound by WBC Ltd, Bridgend, Mid-Glamorgan.

Contents

Acknowledgements

My thanks go to: R. John Griffiths, FBHI, Curator of Prescot Museum, Merseyside, whose early help and encouragement has been invaluable; Rita Shenton, whose explanations and advice on the book business in general has assisted me over many rough spots; Wilf Styles, who produced many amazing discoveries, and Barbara Redican, that patient decipherer of my handwriting, who transferred the book onto floppy disc.

Illustration Credits

All photographs are by D.A. Tennant and those marked with an asterisk below were taken with the kind permission of the organizations mentioned.

Black and white illustrations
Liverpool Museum: 86, 148. National Trust: 101. Old Sturbridge Village, Massachusetts: 104*. NAWCC Museum at Columbia, Pennsylvania: 105*, 109*, 110*, 111*. American Watch and Clock Museum, Bristol, Connecticut: 106*, 107*, 108*. Newark Houses Museum, Leicester: 136*. Merseyside Museums: 150*.

Colour illustrations
National Trust, Dudmaston Hall: *vii*, Merseyside Museums: *xix*

Glossary

Acanthus A stylized serrated leaf motif used in decoration.
Annealing A method of softening metal by heat treatment.
Arabic Quarter Markings Arabic minute numerals at the 15, 30, 45 and 60 places on the dial.
Arbor A steel shaft and pinion to which a wheel is attached.
Armillary Sphere Model of Ptolemy's universe with the earth in the centre with bands showing the equator, tropics etc. Originally used in ancient astronomy and timekeeping.
Automaton Moving figure or other object, such as ship, Father Time, etc., usually operated from the pendulum.
Back Cock A bridge-shaped piece screwed to the back plate and supporting the rear pivot of the pallet arbor and the suspension spring at the top of the pendulum rod.
Back Plate Rectangular brass plate forming the back of the movement.
Barrel Brass drum around which a cable or gut line is wound. It has ratchet teeth on the outside perimeter.
Barrel Arbor The shaft on which the barrel turns.
Bob The usually circular weight at the end of the pendulum rod.
Break Arch Dial Where the central part of the top edge is formed into an arched curve.
Cartouche An ornate frame, often oval, sometimes holding the clock maker's name in the centre.
Chapter or Chapter Ring The annular area of the dial showing the hours and either numbers or marks for the minutes.
Click A spring-loaded pawl, bedding into ratchet teeth.
Craquelure The fine cracking due to shrinkage found on a painted or glazed surface.
Cupping Seen where the edges of paint flakes bend upwards.
Date Dial A subsidiary dial indicating the date seen through an opening in the main dial. It can also be on the main dial in the form of a circle below the centre hole.
Date Ring A silvered brass annulus with the date numbers engraved on it, suspended on rollers or posts on the back of the dial and seen

through a small aperture in the dial.

Dial Feet Three or four pillars, usually brass, on the back of the dial, which fit into the false plate or front plate. They can be short, fitting into the false plate, or long, going directly into the front plate.

Drop Dials or English Dial Clock A wall clock with a round dial.

Ecliptic Apparent path of the sun around the earth.

Escapement File A small fine-toothed file.

False Plate A rectangular sub-frame at the back of the dial made of cast or sheet iron, in order to easily attach the dial to the movement.

Front Plate A rectangular brass plate forming the front of the movement.

Gesso A mixture of plaster and size which was gilded in order to give a raised pattern on a dial.

Grandfather, Longcase or (in USA) Tall Clock Driven by weights and standing 6 ft. or higher.

Grommets Pressed brass linings for the edges of winding holes. Grommets were originally called collets.

Hand Collet A large brass washer placed over the minute hand and held with a taper pin pushed through the hole in the end of the centre arbor.

Lines Gut or cable wound around the barrel and supporting the weight pulleys for driving the clock.

Marriage Where a dial has been fitted to a movement other than the original.

Pallet Arbor The arbor on which the pallets of the escapement are fitted.

Pegwood A fine grain small diameter wooden dowel.

Pillar and Scroll Shelf Clocks Type of rectangular American clock with pillars at the side and a scroll-like pediment.

Pipe and Post Method A post attached to the back of the main dial so that the moon or date dial could revolve on its pipe.

Planishing Hammer A type of hammer with a polished, square, slightly domed end used to flatten metal.

Reticulated A surface divided by a network of open cracks.

Silvering A method of chemically coating brass with silver.

Strike/Silent Dial Usually in the break arch or a corner where a pointer can be moved to make the clock strike or remain silent.

Thirty Hour Clock A clock that has to be wound every day.

Transfer Printing An engraved plate is inked and a special paper takes up the ink, which is put on the dial by use of a roller. Most maps of the world were done like this, and it was also used for decorative elements.

Foreword

There is one word which almost always produces an instant response from most people: 'longcase', 'grandfather' or, in the case of the USA, 'tall clock'. They may possess one or it may have been part of their childhood. They often sadly recollect the disappearance of such an item – almost as if it were a real person. This book concerns the longcase painted dial which revolutionized the appearance of clocks for a period of roughly one hundred years from 1770. These painted dials used to be considered inferior to brass dials, but now they are being truly appreciated, as witnessed by the ever escalating saleroom prices. As far as is known, the birthplace of the painted dial was Birmingham. Many factors combined to facilitate this innovation, but one of the most decisive was the emergence of the canal system which brought fuel and raw materials to the centre of Birmingham. A great number of clockmakers were able to take advantage of this and they began producing movements and components for sale to the trade. Initially these early painted dials were more expensive than brass ones but, as they became more popular, the price dropped.

As far as can be ascertained large clockmaking factories in the early period (pre 1810) did not exist in Birmingham though often several clockmakers and apprentices formed a small manufactory, sometimes using batch production methods. Clockmakers in other parts of the country continued making individual movements, often surprisingly large quantities. They would have had to order their painted dials from Birmingham. Those clockmakers in the city itself were often in close contact with the dial makers. In some cases even in the early period false plates were not used on 8-day clocks which pre-supposes a co-operation between clock and dial makers. From 1800 onwards more and more dial makers' establishments appeared, causing the price to fall even further and in some cases, especially after 1830, the quality deteriorated. The wooden cases were usually made locally and the whole clock was priced according to type, such as: 8-day with moon, arched 8-day, square 8-day or 30-hour. Special orders, such as automata, musical or other complications of course cost more and even the more elaborate corners such as the Four Seasons etc. were marked 'extra price'.

These longcase clocks, which could truly be called ordinary people's clocks, often remained in the same family for many generations, especially in country areas. They are frequently regarded as a somewhat difficult elderly relation needing attention from time to time. Others, unfortunately, suffered dire neglect, with the case used as firewood and the movement sold for scrap or just dumped. Still more were bought up by enterprising antique dealers and shipped abroad by the container load. Painted dials were frequently removed and brass ones substituted, or the whole clock might be turned into a cocktail cabinet.

This book concerns those dials which remain – although quite a few have been separated from their original movements. It is hoped that this book will help to increase the awareness and appreciation of this unique form of art which came into existence during a most interesting era of British history.

For my husband without whom this book
could never have been written.

1 The History of Painted Dials on Longcase Clocks

THE CONSTRUCTION OF LONGCASE PAINTED DIALS AND FALSE PLATES

Before beginning to discuss the history and development of the painted dial in Great Britain it would be useful to know how and of what materials these dials and their attached subsidiary dials and false plates were made. Considerable confusion has been caused by calling these painted dials 'enamel', which is incorrect because the painted process was called Japanning. The term 'enamel' when referring to a painted dial is often found in older books such as *English Domestic Clocks* by Cescinsky and Webster or *The Grandfather Clock* by Ernest L. Edwardes.

The basic material for the main dial was wrought iron, though a very few were made of brass. Painted wooden dials will not be dealt with here, except in the section on Birmingham influences in American clocks. The wrought iron has a distinct texture to it, which can easily be seen on the back, but late dials, usually after 1830, are much smoother so either the metal or its finishing had improved. The gauge of the iron varied enormously. F. Byrne of Birmingham used 0.065″ in 1790, and Finnemore, also of Birmingham, used 0.086″ in 1840.

Before about 1825 all longcase dials were flat, but after that the convex dial appeared, where the round centre section containing the numerals had been pressed to a gentle curved shape. In order to achieve this, the iron plate had to be somewhat thinner than for a flat dial. Not all dials produced after 1825 were convex, as the flat dial was easier to make and the convex one was probably more expensive. Although this book is primarily about longcase dials, a very few bracket clocks will be considered. The iron plates used for these were usually thinner than the longcase ones, and they employed the pressing process earlier, particularly in round dials which might have been inspired by the French vitreous enamel dials.

The shapes and sizes of longcase clock dials varied a great deal as can be seen in the copy of Wilkes and Son Japanned clock dial list. This list does not include the 11″ early square 30-hour dial used by James Wilson of Birmingham in the 1780s, nor the giant 16″ ones of the 1850s, where the dial maker remains unknown. The measurements in the list are those of the metal sheets before painting. Some dial makers used odd sizes like 13½″. Oval dials tended to be rare, and one oblong dial of 1830 has been seen.

Birmingham, January, 1820

PRICES

of

Japanned Clock Dials,

MANUFACTURED BY

Wilkes and Son,

WHITTALL-STREET, BIRMINGHAM.

				£	s.	d.
FOURTEEN inch	Arch	Moon Dial	.. ..	1	4	0
Thirteen inch	ditto	ditto	.. ..	1	2	0
Twelve inch	ditto	ditto	.. ..	1	0	0
Fourteen inch	Square	ditto	.. ..	1	2	0
Thirteen inch	ditto	ditto	.. ..	1	0	0
Twelve inch	ditto	ditto	.. ..	0	18	0
Fourteen inch	Solid Arch	Eight Day	.. ..	0	16	0
Thirteen inch	ditto	ditto	.. ..	0	14	0
Twelve inch	ditto	ditto	.. ..	0	12	0
Fourteen inch	Square	ditto	.. ..	0	14	0
Thirteen inch	ditto	ditto	.. ..	0	12	0
Twelve inch	ditto	ditto	.. ..	0	11	0
Fourteen inch	Round	ditto	.. ..	0	13	0
Thirteen inch	ditto	ditto	.. ..	0	11	0
Twelve inch	ditto	ditto	.. ..	0	10	0
Thirteen inch	Arched	Thirty-hour	.. ..	0	13	0
Twelve inch	ditto	ditto	.. ..	0	11	0
Eleven inch	ditto	ditto	.. ..	0	10	0
Fourteen inch	Square	ditto	.. ..	0	13	0
Thirteen inch	ditto	ditto	.. ..	0	12	0
Twelve inch	ditto	ditto	.. ..	0	10	0
Eleven inch	ditto	ditto	.. ..	0	8	6
Ten inch	ditto	ditto	.. ..	0	7	0
Fourteen inch	Round	ditto	.. ..	0	12	0
Thirteen inch	ditto	ditto	.. ..	0	10	0
Twelve inch	ditto	ditto	.. ..	0	9	0

Six inch	Round	Spring Dial	.. ..	0	7	0
Eight inch	ditto	ditto	.. ..	0	8	0
Six inch	Arch	ditto	.. ..	0	8	6
Eight inch	ditto	ditto	.. ..	0	10	0

Oval and Round Dials of all Sizes.

Additional Paintings and Movements charged as follows:

Seasons, Quarters, Virtues, Elements, &c.	0	8	0
Landscapes or Figure Pieces in Arch	0	4	0
Single Figures painted in Arch	0	2	0
Adam and Eve to move	0	7	6
Ditto, if Serpent to move	0	10	0
Ship to move, Old Time, &c.	0	6	0
Swan's Neck to move	0	6	0
Boy and Girl swinging, Harlequin and Columbine, &c. ..	0	7	0
Shuttlecock and Battledore to move	0	7	0

Movements and Paintings of every Description.

Printed by R. Peart, Bull-Street, Birmingham.

(from the Peter Stubs Archives courtesy of Manchester City Art Galleries)

The separate date dial was supplied with the main dial. They were usually brass, though iron was sometimes used. Often the brass ones had the dial maker's name stamped on them. The manufacture of brass in the early period was a very hit or miss affair which resulted in variable quality. Some particularly early (1770s and early 1780s) date dials appear to have been flattened using a planishing hammer. Most date dials show marks of being rolled, though hammering could have been used as well. Some of the brass had so much copper in it that it appears very red when cleaned.

Moon dials in the early pre-1800 period were usually brass, but later on iron was used, and after 1840 iron moon dials were quite common. One unusual moon dial of 1850 was made of zinc. The thickness of the brass used for moon dials varied enormously from 1.5 mm to 1.05 mm and to the thinnest at 0.85 mm. One particularly thick moon dial was used by Osborne of Birmingham *c.*1795 and exported to Ireland. The teeth on this moon dial had been filed, which is unusual except in Ireland. (See 'The Painted Dial in Ireland' by M.F. Tennant, *Antiquarian Horology*, Summer 1991, pp. 413–18.)

The false plate or subframe was attached to the short stub feet of the main dial in order to allow the clock maker to fit his movement more easily. These short stub feet were usually made of brass but a very few were iron. The early false plates were all grey cast iron, and most of them had the dial maker's name cast in to them. A few early un-named cast iron false plates exist, as well as some which had small numbers on them. The cast iron false plates were made from patterns ranging from the very plain to the elaborate. The original master pattern could have been brass or wood. The false plates were probably made in lots of at least four to a box, and as time went on the master patterns became worn and clogged with sand, which resulted in almost illegible names late in the run. Some false plates have names on both front and back. After about 1800 the name Wilson often appears on both sides, and there are examples of a faint Wilson on one side and a sharp W. Francis, Birmingham on the other. Invariably the sharper name is the real one. It could have been possible that the old masters were re-used with a new name on the other side. Dials usually fitted only the false plates that were supplied with them, and even those from the same maker were not always interchangeable, as some antique dealers have found out. However, just to make things more confusing, a few late un-named Birmingham dial makers used up old false plates, usually Wilson ones, and attached them to their new dials. A variety of cast iron false plates can be seen in **Illustrations 1–4**.

Cast iron is a very brittle material and fractures very easily, especially when dropped on a stone workshop floor. Quite a few cracked false plates have early clumsy repairs to them. This fragility and the cost of casting may have been one of the reasons for the switch over to sheet metal false plates around 1825–30. These sheet iron false plates were often stamped with the dial maker's name, but late ones, after about 1850, are generally blank. Very rarely a cast iron false plate belonging to a pre-1800 dial has been replaced by a sheet iron one, either because of damage to the original one or for some other unknown reason, but the paint on the new false plate usually does not match that on the back of the dial. Once the size and shape of the main dial had been decided on, the apertures had to be cut out. These included break arch moon, automata in the break arch, 12 o'clock moon, date aperture, or hole for date hand, centre hole and hole for seconds hand. Sometimes the winding holes were made at the same time, but they also could be made later. It is easy to see if a hole was made before the dial was painted because there are paint runs on the back. Very rarely mistakes were made or an order was changed so that a hole cut for a seconds hand was not needed, and this was plugged with lead before painting.

After the holes and apertures were cut the brass feet were riveted in. Generally 8-day dials had four short stub feet to fit with false plates, and 30-hour dials had three long feet and no false plate, but there are many 8-day dials with three or four long feet and no false plate, and some 30-hour dials had false plates in order to accommodate false winding squares which imitated the more expensive 8-day clock. Any posts or other additions such as rollers for a silvered brass date ring would have been put on at the same time as the feet were riveted in. However, some

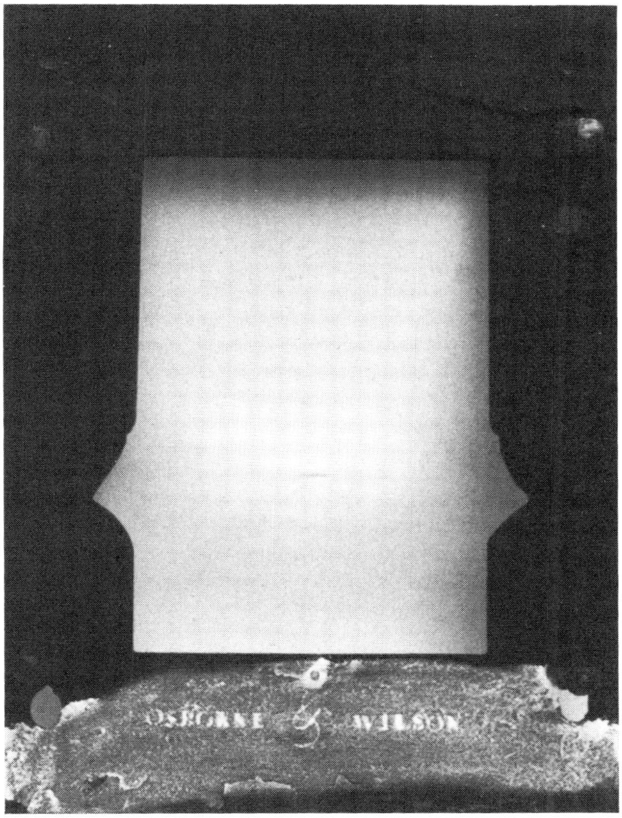

1 Very early (1772–7) false plate made for or by Osborne and Wilson of Birmingham, the originators of the painted dial and the false plate system. The dial belonging to this false plate is shown on pages 29 and 217

clock makers put in various pins after the dial was painted. The locating of the feet on the dial was not always standardized, though an attempt at this was made by Samuel Harlow of Ashbourne in his 'Clock Makers Guide' of 1813, which laid out plans for both 8-day and 30-hour clocks. After 1813 the false plate pattern included in the 'Guide' was used by some makers, but certainly not all of them followed this plan. Some dials show centre punch marks on the back which located the positions for marking out the arch. Wilson dials often had a Maltese Cross punched on the back, and other dial makers had their whole name on the dial, or else just a number.

However well the dial feet were riveted in, a small depression was usually left. When an old dial has to be stripped down to the bare metal because of extreme damage such as fire, a black or brown sealing wax type of filler can be seen in these places. This filler was put on before any primary coat. Iron plate is notoriously prone to rust, so a primary coat of red iron oxide paint was used on the back and front, as well as on the false plate. Iron oxide is an ideal primer, as it serves as a good key for any subsequent paint layers. However, for some reason or other different types of primer were sometimes used, often with disastrous consequences. A pre-1800 Manchester made dial had a primer that included a bituminous compound that never really hardened even after 200 years. Another 1850 dial had a dark red shiny primer that provided no key at all for the subsequent painting, and a number of late 1850s

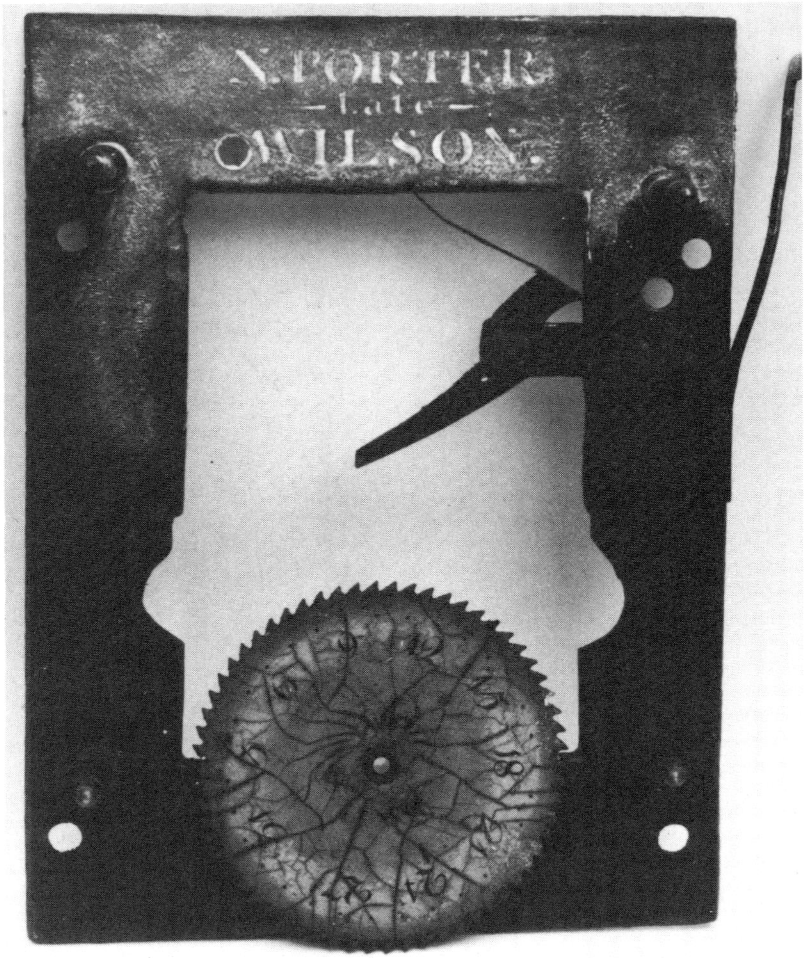

2 Rare false plate from the short-lived successor to Wilson, Nathaniel Porter (1809–12). This false plate has 'Wilson' on the reverse and there is a date dial and mechanism attached. For the main dial see Illustration No. 44

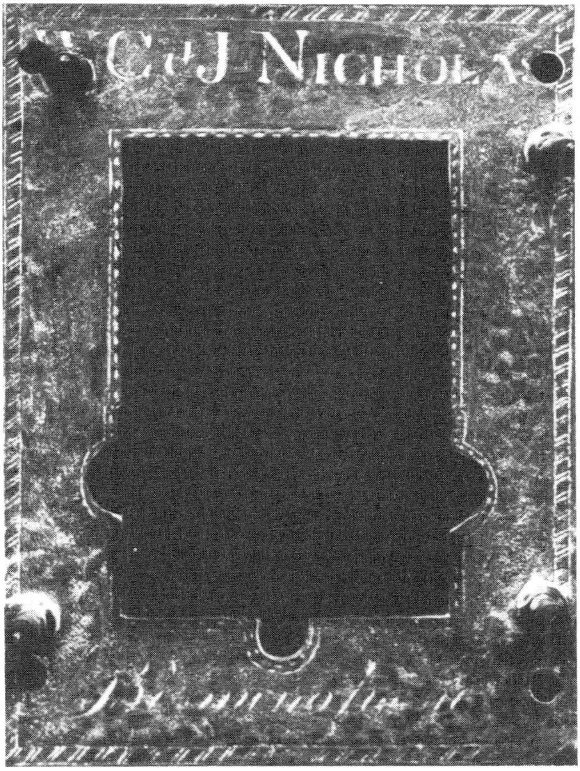

3 Elaborate false plate made by William, Caleb and Joshua Nicholas of Birmingham (1787–1825) attached to dial (Illustration No. 40). See page 75

dials used a shiny black very brittle primer which also did not bond with the paint. The back of the dial sometimes was painted over the primer in black, brown black or red brown.

The front of the dial was painted with ground paint, sometimes as often as six times in pre-1800 dials. This provided the enamel-like surface that was so admired. This ground or base paint was analysed by emission spectrometer and electron scanning. It showed, as expected, a large amount of lead, and a considerable quantity of aluminium and silicon. Analysis by electron diffraction showed the lead to be basic lead carbonate. The aluminium and silicon were probably a form of clay filler. The proportions and colour of this paint varied enormously. The colouring matter was not as finely ground as in modern paints, and on close examination distinct specks can be seen. Sometimes the filler was not evenly mixed, which led to small round pale patches, though this may not have been visible at the time of painting. The base paint was put on with a brush. An early Osborne dial showed distinct brush marks. Sometimes the back of a dial showed circles where a paint pot had rested, and occasionally the paint on the back held a brush hair. The base paint was dried with gentle heat, not a true firing as in vitreous enamel. The dial was rubbed down with an abrasive between each coat of paint in order to make a flat surface until the last and most perfect layer was put on. Later, around 1830, dials tended to have fewer and

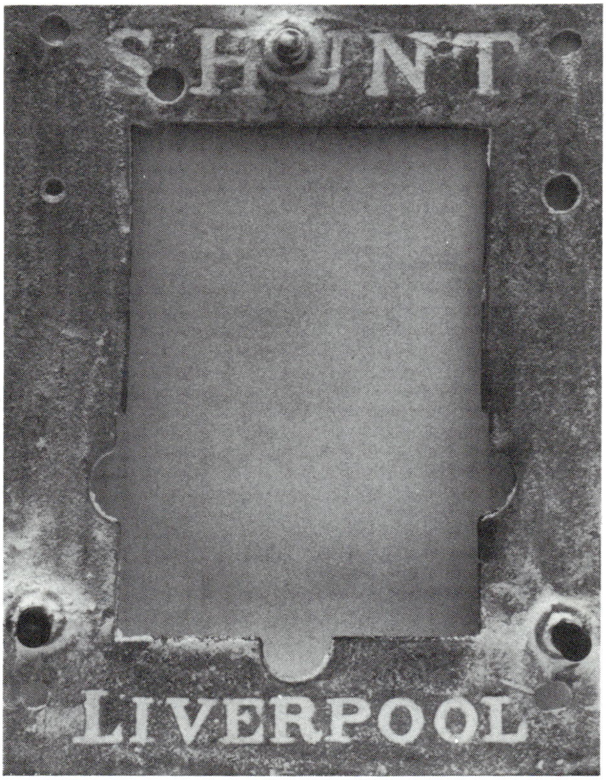

4 A Samuel Hunt of Liverpool false plate *c.* 1845 belonging to dial (in Illustration No. 101). See page 155

thinner layers of paint, and the convex dials often show uneven paint application. The date and moon dials made of brass had no priming coat, and the date dial was of course painted to match the main dial. Iron moon dials often had a black priming coat. Most of the brass moon dials seem to have had an off-white layer of paint applied first.

It seems obvious that a template of some sort was used in planning the graphics part of the dial. Quite a lot of dials show the pricking-out marks for minute dots and those for the tops and bottoms of Roman numerals as well as seconds and date dots if these were to be included. Sometimes the pricking-out marks were not exactly followed, especially on the Roman numerals, and mistakes on seconds markings did happen. A scriber could have been used to prick out the markings, but date and seconds dots often appear to have been done with something that had a distinctly sharp, square end.

The concentric circles were done with a compass using India ink, and the straight lines were made with an aligning pen. The Arabic numerals and the signature were done with either a brush or a quill pen, using the same India ink. Very rarely on late (post-1840) dials a steel nibbed pen was used for the signature, and this tended to dig into the paint. Filling in of the Roman numerals and parts of the Arabics was done with a brush using either ink or lampblack oil paint.

The places for the artwork were roughed in, using a lead pencil which can sometimes be seen through the artist's very thin oil colour. In general, only the fruit and flower motifs were thickly painted, almost giving a relief effect. When the artwork was dry, it was varnished. In the early period this was done most carefully, following the motifs so that it could hardly be noticed. Later on dial workers became more careless or hasty so that a lot of overlapping could be seen, especially now that the varnish used has darkened, sometimes giving a 'transfer outline' to the work.

If there were maps of the world covering the so called humps of a break arch or 12 o'clock moon, these maps would have been put on using a process from the china industry called transfer printing. A copper plate was engraved with the maps, inked and pressed on to a special paper which was then pressed down with a roller onto the dial. Different sizes of dials needed different engravings, and in the early pre-1800 period each dial manufacturer usually had his own set of plates. Sometimes the maps were updated, especially in the case of Wilson and Osborne. Occasionally mistakes were made in this process. The paper could have had a wrinkle in it, or else the maps were put on upside down as in the case of an Osborne dial with a 12 o'clock moon. Very rarely the usual position of Wilson maps was reversed so that instead of the Americas being on the left, they were put on the right. The ink used was of a different type from that of the numerals.

Quite a lot of early dials used a type of plaster called gesso for raised work which, when dry, was gilded using gold leaf. This raised work was often found framing corner motifs, or decorating the whole corner in a pattern, or else in quite elaborate break arch urns and scrolls. It was also used in framing oval scenes on the break arch, and a few makers scattered small gesso flowers inside the chapter ring, often at the places where the feet had been rivetted in. The adhesive used in any gilding work was Japan gold size, which is a pale straw colour when new, but after 200 years it can darken to a deep umber tone in places. Sometimes lampblack appeared to be mixed with the gold size giving a pale grey colour, but whether this was done to make the size more visible or just happened through contamination is unknown. This grey colour only seemed to turn up on post-1830 dials with a gold band on the outside of the chapter ring.

Flat gold leaf could have been used almost anywhere on a dial. Most usual was the previously mentioned gold band on the outside of the chapter ring, but when a date aperture was used the edges of this were nearly always gilded. Gold leaf could form the background for a corner design or be overlaid with translucent colour when forming part of a lady's dress on those dials showing the four seasons. From about 1840 silver leaf was also used, serving as a substitute for the more expensive gold leaf. Possibly tin leaf was used too.

Moon dials usually had the circles for the moon faces scribed into the dials. The faces and scenes were done first, and then the black or very dark brown 'edging' was painted on using lots of varnish mixed with the paint to hasten drying. Some very early moon dials used gold leaf stars and even a comet on a deep blue background, which is similar to those on brass dialled clocks. Moon dials were usually quite heavily varnished, sometimes using a copal varnish. Automata, where there were moving parts, could have been made of thin brass or iron, but they had to be as light as possible so as not to take too much energy from the mechanism. The automata were also painted with artists' oil paint.

WHERE TO FIND PAINTED DIALS

Where does one find painted dials? The answer is, of course, just about everywhere, but certain places are more likely to have them than others. The serious student should ask to see the reserve collections in museums.

In general, there still seem to exist more painted dials in the Midlands, the North, Scotland, Wales and some areas of the West Country, where these dials were originally the most popular.

Reasons for Classifying Dials

There are many reasons for attempting to classify this uniquely British art form. These dials were produced in large numbers during a period of roughly one hundred years, from 1770 to 1870, and they were exported all over the world, especially to the United States. Clocks with painted dials are still being exported by the antique trade. Probably about 90% of painted dials were made in Birmingham, but there were a surprising number of other production centres.

One of the most important reasons for trying to classify painted dials is the question of trying to date the object. Of course most dials have a clock maker's name on them which can be looked up in various dictionaries, but some clock makers produced clocks for fifty years or more, and a number of names remain unlisted. Also, the name now visible on the dial may not be the correct original one. Furniture experts could look at the clock case, but there are a great number of caseless dials, as well as dials in cases that obviously did not belong to them originally. As well as putting a date on a dial, it is useful to know just where it was made. This is quite easy where a date dial or false plate plainly says Birmingham on it, but it is not so simple when things are unmarked. As in most branches of art, there is an orderly progression of styles and graphics that can help this process of classification. Not every dial exactly fits all the rules, because customers' and clock makers' requirements varied greatly.

Luckily, some of the correspondence between a clock maker and his factor who supplied the dials in Birmingham still exists (see the Deacon papers). When one knows what a dial of a particular period and maker ought to look like, then it is easy to detect spurious 'improvements'. One thing that can confuse accurate dating is the fact that some dials were completely repainted by a different dial maker to the original one in an effort to modernize the dial, or else a lot of blank dials and false plates were bought up when a manufacturer went out of business. Usually this can be detected because the dials in question are absolutely different in graphics and decoration to anything that particular maker produced before. Also, the clock maker's dates do not agree with the dates of the manufacturer, though this does happen sometimes when a clock maker may not have been accurately listed in one of the dictionaries or in some of the books on clocks of a particular locality.

There is now a tremendous interest shown by the general public in those clocks with painted dials, especially when one can explain just how and when a family heirloom was made. Years ago it was sometimes hard to even give these clocks away, and many had the cases burnt up for firewood and the mechanisms sold for scrap, leaving the dial to rust away in a shed. Now the clocks with painted dials are beginning to take their rightful place as desirable objects. There is also an increasing interest in local history, and an accurate knowledge of a particular dial may assist the local historian in tracing a previously unknown clock maker or else one whose dates and places of work were not defined.

2 How the Painted Dial was Developed

EARLY ROOTS

The development of the painted dial is an extremely interesting one. Paint has been used on clock dials from the sixteenth century gothic clocks through the Dutch Stoeltjeklok to the beautiful bracket clocks made by Stephen Rimbault in the 1760s where the top of the dial often contained automata painted by the famous artist Zoffany. However, brass dials were used by all British clock makers up to the advent of the painted dial in the 1770s with one important exception. True vitreous enamel longcase and bracket clock dials were made from the 1750s to the 1770s, but these must, in the longcase dials at least, have been incredibly difficult and expensive to make. There do not seem to be a great number of these real enamel longcase dials about.

True enamel work was very popular during the eighteenth century, and these vitreous enamel dials which remained so brilliant were a real change from the brass and silvered clock dial that could soon become dark and tarnished in the often smoky atmosphere. It seems obvious that the vitreous enamel dials paved the way for the painted dial that came so soon afterwards. Anthony Theelke in his well-illustrated booklet 'Faces of Mystery' deals with a number of these attractive dials. He also illustrates designs of fantastic birds, flowers and ornamental scrolls from a book by Robert Sayer, *The Ladies Amusement*, published in 1758. This book had a great influence on many artists and craftsmen for decades, and particularly the birds on the enamel dials seem to be directly copied from the fine engravings. This same book could have inspired some of the decorative work seen on the painted dial.

Brass dials, especially in the North, became increasingly elaborate in the centre engraving and sometime in the 1770s the fully painted centre appeared. This matched in style the brass moon dial, and could be put on a dial with a break arch moon of the 12 o'clock type. There do not seem to be a great many of these dials still about, but those that have survived came from around the Manchester area. Another quite unique precursor of the painted dial was a brass dial with an engraved centre and fully painted corners that extended into a break arch complete with moon. This dial was made for William Kirk of Stockport *c.* 1765 and can be seen at the Lady Leverhulme Museum, Port Sunlight, Merseyside.

There was another development in dial making which probably started in about 1770, and that was the one-piece silvered engraved dial. These one-piece dials were very popular in some areas, especially Scotland, but they depended on a very good engraver and the silvered surface often quickly became tarnished. See Felix Hudson, 'Longcase Engraved Dials in mid Scotland', *Antiquarian Horology*, Winter 1980, pp. 392–408.)

OSBORNE AND WILSON PARTNERSHIP 1772–1777

The first book to deal exclusively with painted dial clocks was Brian Loomes's *The White Dial Clock*, published in 1974. The second edition, *White Dial Clocks – The Complete Guide*, published in 1981, had a great deal of additional information, including a much more complete list of dial makers. Many earlier books on clocks include a section on White Dials. Eric Bruton's book *The Longcase Clock*, first published in 1964, appears to be the first book to take painted dials seriously. One very useful book that deals with just Birmingham dial makers and clock makers is Joseph McKenna's *Watch, Clock and Dial Makers of Birmingham 1547–1900*. There are a number of clock books devoted to a particular area, and a list of these can be found in the bibliography.

On 21 September 1772, there appeared in *Aris's Birmingham Gazette* an advertisement that would completely change the appearance of British clocks.

> **White Clock Dials**
> Osborne and Wilson, manufacturers of White Clock Dials in imitation of enamel, in a manner entirely new, have opened a warehouse at number 3 Colmore Row, Birmingham, where they have an assortment of the above mentioned goods. Those who favour them with their orders may depend on their being executed with the utmost punctuality and expedition. NB: The dial feet will be riveted in the dials and such methods used as will enable the clock makers to fix them to the movements.

The influence of the true enamel ware is noted in the advertisement, and early dials certainly bear remarkable decorative resemblance to enamel pieces of the period, such as snuff boxes, candlesticks, needle cases and other articles.

As a matter of possible interest, a street map of Colmore Row *c.* 1850 shows No.3 to have been a building with quite a wide frontage and a long back plot which had access to a short lane off Livery St. A photograph c. 1870 reveals the building to be of 3 storeys, probably with a basement area.

Thomas Hadley Osborne, the Osborne of the advertisement, was related to the clock making family Hadley of Birmingham. Before embarking on their venture, Osborne and Wilson must have made a number of trial pieces, but unless one is very lucky the chances of finding an authenticated one of these pre-September 1722 dials is slim. Susan E. Stuart in her article 'First White Dial Longcase Clocks and their Cases' (*Antiquarian Horology*, December 1982, p. 534) mentions a notation in the *Waring and Gillows Waste Book*, dated January 1772, referring to two 'enamelled faces'. Whether this was true vitreous enamel or a painted dial is not certain. However, a definite mention of a 'Japanned or enamelled dial' was made in May 1772, which shows the two terms to be interchangeable. There is, in that article, a photograph on p. 542 of a painted dial made for Thomas Worswick where the raised corner decoration is similar to that of a dial illustrated on p. 206 of Brian Loomes's book *Grandfather Clocks and their Cases*. That particular dial had a break arch moon, and above the numerals for the moon's age there was a painted 'Osborne and Wilson Birmingham' which shows that the dial was an early trade sample. These extremely early Osborne

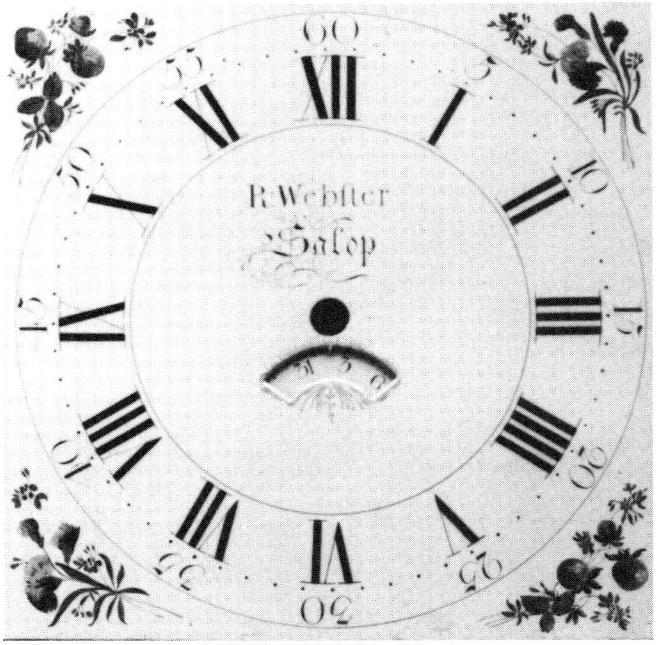

5 Very early (1772) Osborne and Wilson dial for 30-hour clock in dated oak case. This clock has been in the same family since it was bought. See pages 29 and 195–201

and Wilson gesso corners are quite distinctive, being bolder and simpler than later productions.

Probably the earliest Osborne and Wilson dial recorded in this book is a simple 30-hour dial made for Robert Webster of Salop (1755–1832) (**Illustration No. 5**). The country-made oak case had a carved inscription on the hood: 'IE 1772'.

This clock had been in the same family ever since it was made, and was bought in Shrewsbury where the owner sold butter at the market there. The dial shows the simple fruit and floral corners, unadorned by gesso. The date aperture is the early type with squared-off ends and a very sharp pointer in the middle. The date dial was not stamped. There must have been a considerable demand for this relatively cheaper form of dial, though when first introduced painted dials were said to cost

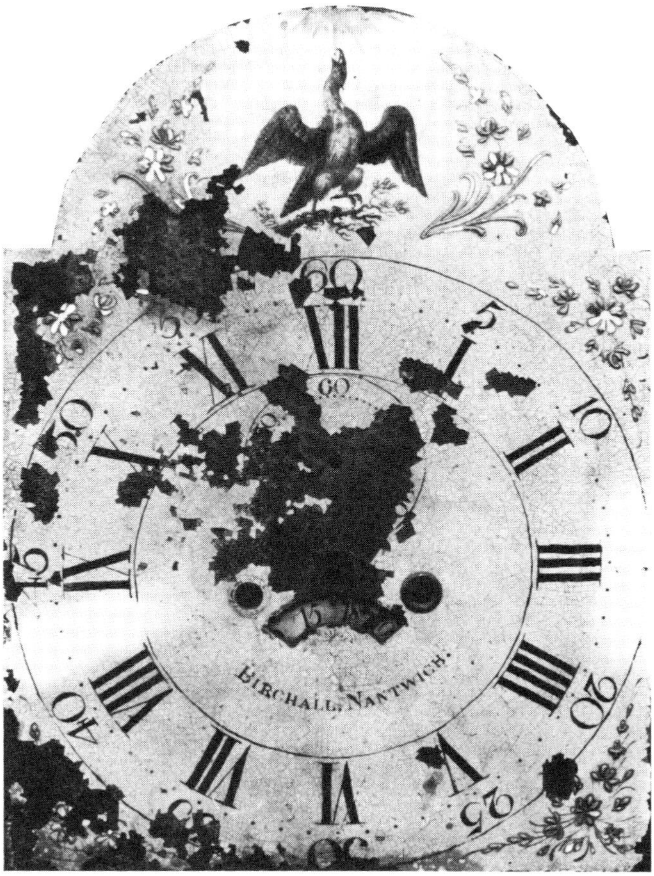

6 Original condition of Osborne and Wilson dial (1772–7) made for Thomas Birchall of Nantwich mentioned in text page 30. See Illustration No. 1 for false plate and No. 139 for restored dial

more than the brass ones. Unfortunately, this dial had to be heavily restored as it was in a positively dangerous state when first seen (**Illustration No. 123**). As a matter of interest, the same clock maker used a similar dial about fifteen years later, but that dial had a rounded date aperture and a centre painting of some auriculas.

Another very early dial was a sorry looking object which turned out to have a nice sharp Osborne and Wilson false plate (**Illustration No. 1**) and that must have come from the period 1772–7. The dial (**Illustration No. 6**) was made for Thomas Birchall of Nantwich (1755–1812). It is sad to see a fine dial reduced to this condition, but its treatment and final results will be dealt with in the section on restoration (**Illustration No. 139**). The rather fierce bird was also featured on a brass-dialled clock that Birchall made which is illustrated in Ernest L. Edwardes's book *The Grandfather Clock* (plate 13). This painted dial had the same early type of date aperture as seen on the Webster one, but the large unstamped date dial had a central hole that fitted on a very short wide post on the main dial with a pin through the post in order to secure the date dial. This type of date dial attachment was soon superseded by the more conventional pipe and post method, but it was used by an early Manchester dial maker. Another sure sign of a very early dial was the presence of gilded winding holes, though this dial had been fitted at a later date with a pair of brass grommets, one of which has been removed for the photograph.

The partnership between Osborne and Wilson only lasted five years, splitting up at the end of September 1777. They must have moved from Colmore Row to 11 Great Charles Street shortly before this, and they appointed a Mr John Walker of Colmore Row to deal with all the legalities. The two streets were quite close together, and it is interesting to see how nearly all the subsequent dial-making firms tended to be in the same area. James Wilson carried on his trade at Great Charles Street while Osborne moved to 20 Cherry Orchard, where he continued working until his death in 1779. The street, Cherry Orchard, no longer exists, but many of the other streets where dial makers worked are still there.

JAMES WILSON DIALS

James Wilson (1777–1809) can be considered the most prolific of all dial makers. Although nothing is known for sure about the organization of his manufactory, he certainly must have developed an orderly system to enable thousands and thousands of dials to be produced in thirty-two years. This restorer has worked on over a hundred Wilson dials in fifteen years, and many more have been examined. This is more than can be said of any other single dial maker, and Wilson dials are found all over the world.

The most important clue as to how the manufactory was organized lies in a little label or job ticket sometimes found on the back of Wilson dials, often in the middle of a white paint splodge. They were printed in blue ink, with a detailed border, and the information was written in black ink, using a fine copperplate hand. Some words were deleted using the same ink (**Illustration No. 7**). The example illustrated came from a dial made about 1795. It is amazing that such a small scrap of paper should have survived nearly 200 years. Of course the things noted on the labels

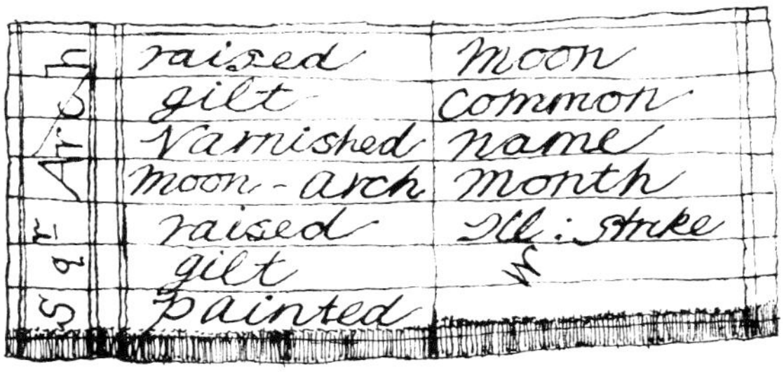

7 Facsimile of job ticket found on the reverse of Wilson dial *c.* 1795. See page 31

varied, and special orders or unusual dials with automata did not seem to carry the labels.

Another interesting point is the presence of black letters or other marks found on the dial back, usually where the white splodge had been put on. These marks were made by the different dial artists. They were not meant to denote the year of manufacture, and many dials have no marks at all, possibly because they were done in ink which can be washed off easily. A, B, D6, G, H̲̅, K̲̅, q, r, W̶, ·:, ⁝ have been noted, and all on one dial; 641, 669, Harlo⊢. Another dial had the ⊢ reversed and the inscription 'Dean 346' as well as the remains of a job ticket. The last mark, the three spots, is most interesting because it can be dated. On removing the bridge of a clock where the dial was marked 'Wooley Tenterden', this inscription was found: 'This clock was made by WS AD 1797'. Thus, it can be deduced that 'three spots' was working at that date. Several more dials with three spots have turned up, and his work seems to have been characterized by very detailed paintings of carnations, though not all of his dials display these flowers.

As previously mentioned, there were many different processes involved in making a painted dial. These processes must have been divided up, with each man having his own job. There were a few women listed as dial painters, but the majority seem to have been men. There must have been separate quarters for some of the work. For example, the sanding down of the base paint created a lot of dust which could not be allowed to contaminate any fresh paint. There appear to have been different artists for the main dials and the moon dials. The application of gesso was also rather messy, and when gold leaf was subsequently laid on it had a habit of blowing about. The basic Wilson gesso pattern · · ⌒ · · ⌣ could vary quite a bit as to size or height and, in general, earlier dials had neater work. Probably the process of transfer rolling on the maps was also a separate department which would take care of those dials with transfer pattern decorations in corners and/or break arches. Thus the workers would have to include graphics, artist main dial, gesso work, gilding, transfer rolling and artist moon dial.

Some dials were sent out with no name or town on them at all, especially those despatched to the United States where it proved in later years to be a great temptation to add a well known name such as Willard. It has been suggested that if more than a certain number of dials were purchased the name would be put on free. Sometimes mistakes in graphics were made, and certainly the spelling of names or places was anything but standardized. For example, the name 'Chaplin' was spelt 'Shoplin' on one Wilson dial and 'Pool' usually meant 'Welshpool'.

In spite of the large number of still extant Wilson dials, no two of them were ever identical. Variations existed even in those dials decorated by the same artist. The graphics tended to be individual, and the moon dials could also be very distinctive, though the moon faces painted by a particular artist certainly had a family resemblance. Those dials with gilded gesso corner decorations, usually ovals and scrolls, were based on a particular pattern, but difference in execution did exist.

The cast iron false plate attached to the dial was an important part of the Wilson production. It is not known whether Wilson had his own foundry for casting these plates, or if he used a firm of iron founders. In any case, Steelhouse Lane, denoting this sort of work, was very near both Colmore Row and Great Charles Street. Most probably the wrought iron plates that made the base for the main dials were produced in Steelhouse Lane.

Once the dials were finished and thoroughly dry, they could have been sold either direct to clock makers or to factors who acted as middle men to tap a more distant market. Clock makers did not usually keep a large stock of ready-made clocks in hand, because the customer generally ordered what he or she wanted. However, there were certainly some examples of painted dials in the shop in order to help the customer choose. Although in the very beginning the painted dial was more expensive than the brass one, with an increase in production the price soon fell. The customers could have anything they wanted put on the dial, but these special orders were always 'extra price', and probably took longer to do.

James Wilson was the most important of all the dial makers. What follows is a survey of the many different types of dials that he made, together with the various changes in style through the years. Other dial makers roughly followed this progression, but no one else provided so many examples.

Any person interested in flowers would be amazed at the list of recognizable species seen on Wilson dials:

Anemone	Lily	Rose (single)
Auricula	Nigella	Rose (moss)
Carnation	Peony	Small vetch
Chrysanthemum	Poppy	Sweetpea (old type)
Crocus	Primrose	Tulip
Dianthus	Ranunculus	Blossom (apple)
Honeysuckle	Rose (double)	

Only two types of fruit were seen: the strawberry and cherry. Birds which appeared on many dials varied enormously and were of no recognizable species. They could be classified as thin birds, medium birds and fat birds.

The corners sometimes had gesso framing them, but it was not always used. After 1800 Wilson dials sometimes used flat gold leaf in the same dot dash pattern, but it was outlined so as to give the appearance of raised work. This same treatment even extended to the corner spandrel-like decorations and those found in the break arch. It obviously cost less to make a dial without gesso work. Another decorative idea was the use of transfer patterns done by the same method as the maps, but these were fanciful scrolls and cartouches for the corners and break arches. The figures of the four seasons, usually feminine, were very popular on the more expensive dials, and the current fashions were well illustrated.

Square dials were the most common, as well as being the cheapest. **Illustration No. 8** shows a square dial made for one of the famous Deacon family clock makers in Barton. It was probably Samuel Deacon, although all the Deacons used Wilson dials. It is a typical 30-hour dial, with its fat bird in the centre and strawberries and dianthus in the corners. The shape of the date aperture had changed from the earlier square ended one, and the date of this dial is *c.* 1790.

Samuel Collier of Eccles (1750–1806) used a dial (**Illustration No. 9**) with fine transfer printing in the centre. This decorated ribbon was more usually done by painting and was used well after 1800. The long necked birds were very similar to those painted by Osborne. The date of this dial is sometime in the 1780s, and because it was used for an 8-day clock it was slightly larger than the previous 30-hour example, and probably more expensive.

It is sometimes thought that the single-handed clock ceased to be

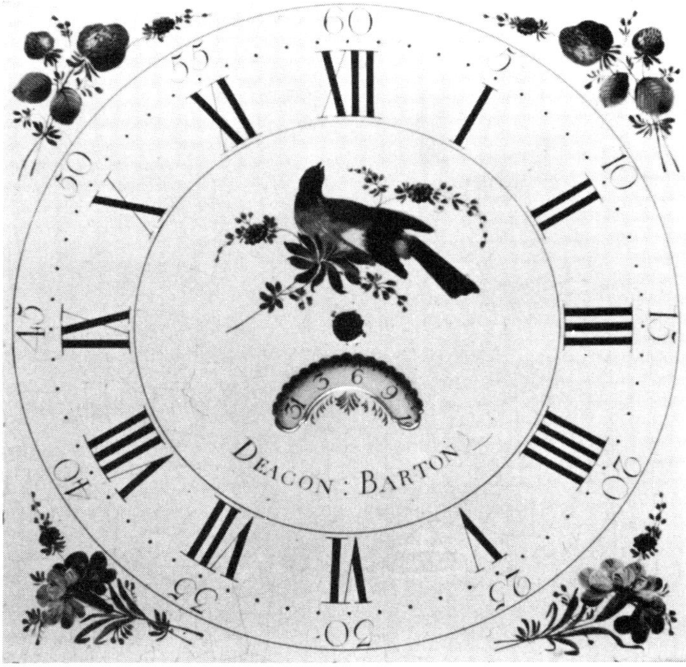

8 Wilson dial for 30-hour clock made for Samuel Deacon of Barton *c.* 1790. See page 33

made in the painted dial era, but Wilson certainly did produce these if required. **Illustration No. 10** shows an unrestored dial made for Samuel and Lazarus of Carmarthen. Its dates are unknown, but the dial appears to be *c.* 1790. The individual centre minute markings can be seen clearly, but the name and town are hardly visible. Unfortunately, there is no movement to this dial and there are no visible markings on the back.

The gesso corner remained popular from the very beginning in the 1770s well into the 1800s, but the designs did vary. The dial in **Illustration No. 11** was made for William Rider of Welshpool, and the corners are slightly reminiscent of the brass spandrels of the so called 'Chippendale style'. The only date given for William Rider is 1792, but this dial looks more like the 1780s. It has the remains of a job ticket on the back, though badly discoloured and impossible to read, as well as the letter 'q'. The date dial had not been stamped.

Just prior to 1800 several interesting style changes in decoration began to appear. The first new idea was the inclusion of scenes, shells or ships in the dial centre. The other novelty was the fan corner. This could be used alone, but often it included tiny vignettes as in **Illustration No.12**, which was made for Catterall of Chorley *c.* 1800. These vignettes usually were done in what appears to be crimson lake,

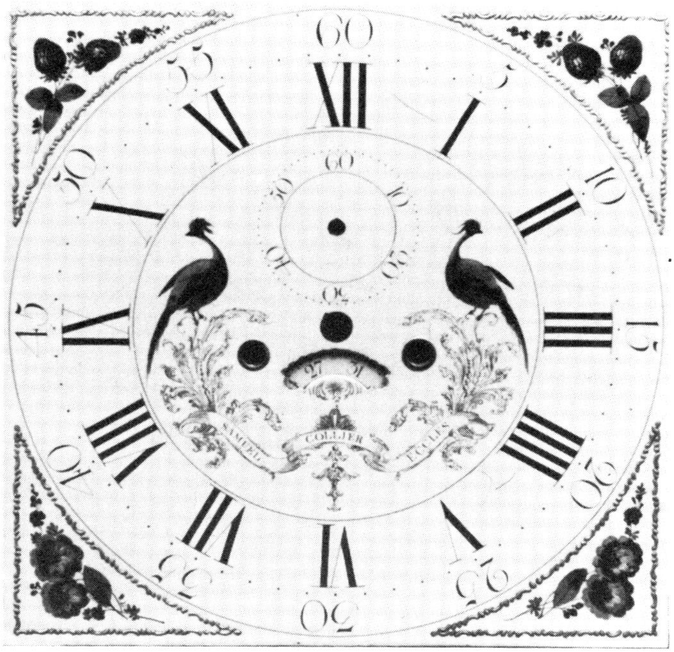

9 Wilson dial with transfer printing in the centre, made for Samuel Collier of Eccles *c.* 1780. This was a much more expensive dial than the one made for Deacon in Illustration No. 8. See page 33

and they could be round, oval or more rarely a sort of diamond shape with the ends trimmed off. In some Wilson dials the round vignettes alone feature in the corners, but they were larger and had a rather heavy dark border or a gold band around them. (An example of this can be found on the dust jacket of Brian Loomes's *The White Dial Clock – The Complete Guide*.) Just before 1800 the style of the graphics changed from Roman numerals for the hours and Arabic for the minutes to all Arabic numerals. The earliest ones had full double Arabic numerals, but later on the minute numbering was reduced to the quarters only with various motifs such as the 'snowflake' or 'fleur de lys' in between. Also, the size of the Arabic quarter markings seemed to diminish and the hour numerals became larger. This change of graphics was a gradual thing, and the clock maker could specify which type of numerals he wanted.

Sometimes a dial appears that does not look like a Wilson one at all yet has the Wilson false plate, stamped date dial and white splodge on the back. **Illustration No. 13** could have been one of the last Wilson ones made before his death in 1809. The graphics have gone back to the Roman type, with Arabic quarters, but the decoration seems untypically Wilson. However, the castles or abbeys painted in the centre were used

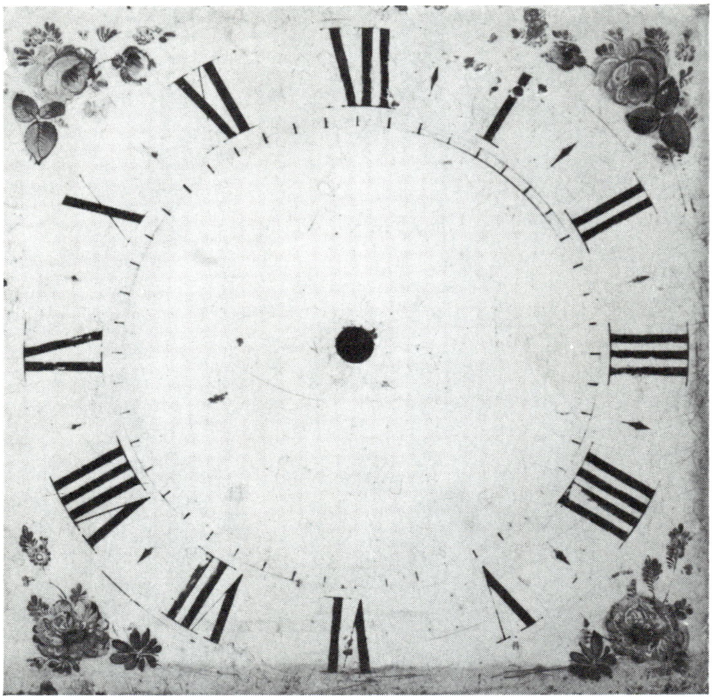

10 Unrestored Wilson dial for single-handed clock made for Samuel and Lazarus of Carmarthen whose date is not listed. See page 34

on another dial with roses and anemones in the corners made about 1803, though that dial had the old full type of graphics. William Holland of Chester (1782–1828) obviously purchased the very latest style, and this clock can be seen at the Grosvenor Museum in Chester. Although Wilson generally used false plates for all 8-day clocks, after 1800 a few dials had four long feet.

The break arch or arched top dial cost a bit more than the square one. There were two sorts of arched top dial: the high and the low one, though some seem to be in between. On some dials the centre punch mark for the placing of the arch can be plainly seen on the back. A variety of subjects could have been put into this arch space, including the following:

flowers
birds
cartouche with name
transfer printing
scenes
scenes with figures
figures alone
gesso alone
gesso in combination with any of the former

One of the earliest of this type is the magnificent example from the National Trust (**Illustration No. 14**) with the gesso work more elaborate than that of the Osborne and Wilson Birchall dial, but using the same type of square ended date aperture. The winding holes are gilded, and there was no false plate, only four long feet. The date dial was attached to the main dial by a short stub post with a pin. The clock maker, John Levich or Levick of Market Raison, was not listed, but the dial gives every appearance of being in the 1770s.

William Toleman of Caernarvon (1791–1844) produced a number of clocks, many of which had Wilson dials. This example (**Illustration No. 15**) was probably made in the 1790s and had one of the most elaborate gesso designs ever seen in an arched top. The urn with floral swags was often used, but it was simpler in the 1780s. This dial had the Wilson Maltese cross stamped on the back.

Transfer printing (**Illustration No. 16**) was fine when it was new, but because the ink was easily water soluble this type of decoration tended to get washed off when someone attempted to clean the dial. The dial was made for Timothy Mason of Chesterfield (1780–91), and the break arch subject was an armillary sphere with garlands. One other such dial exists somewhere, but with very little of the original decoration left, as far as the break arch is concerned.

One particular lady often was featured in the break arch of Wilson

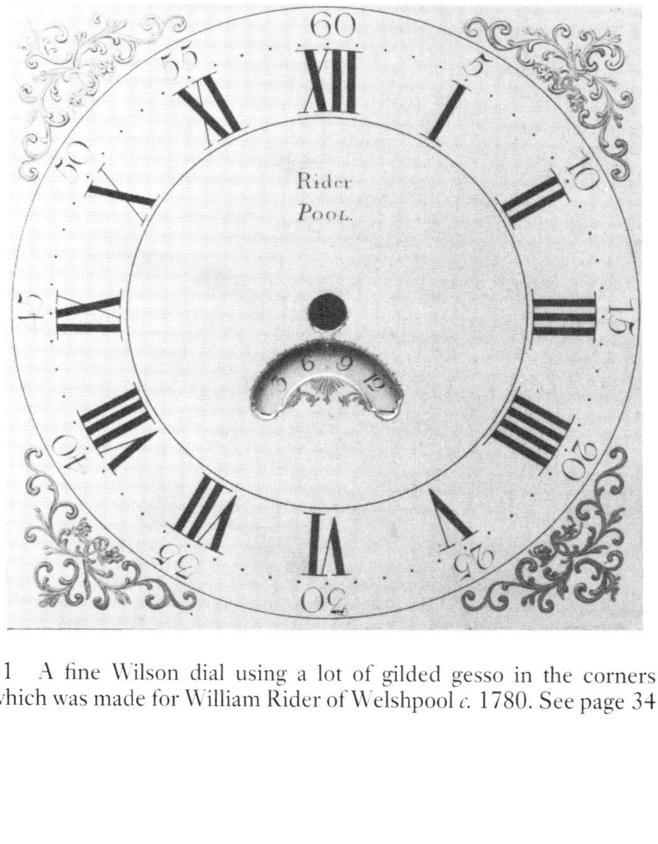

11 A fine Wilson dial using a lot of gilded gesso in the corners which was made for William Rider of Welshpool *c.* 1780. See page 34

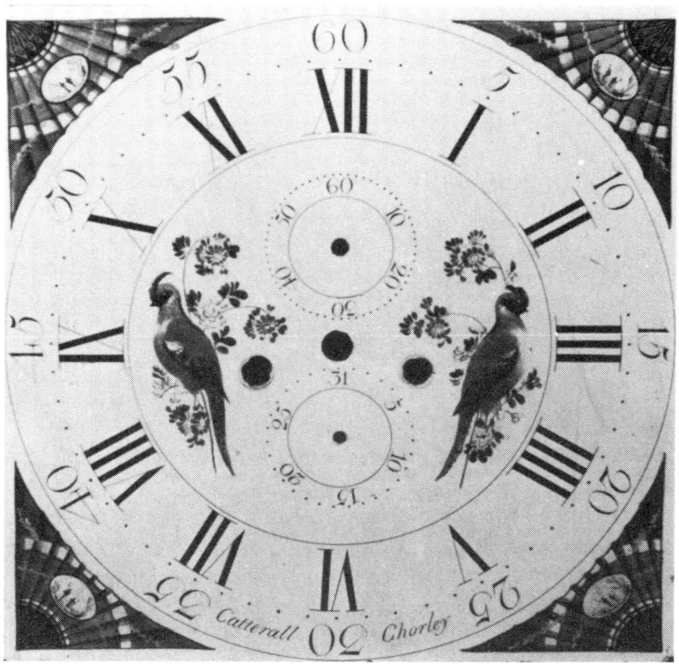

12 New type of fan corners on a Wilson dial with tiny vignette superimposed on the fan. It was made for Catterall of Chorley *c.* 1800. See page 34

dials. She could be seen standing or seated by an urn (**Illustration No. 17**). It is an example of a 'low oval', which meant the scene had a horizontal axis. Usually the scene was framed in black or very dark brown with a few gesso decorations incorporated. This dial was made for T. Lumsden of Newcastle about 1780, and it had the letter 'G' on the back.

Special orders are always interesting, and this was obviously one of these. **Illustration No. 18** shows the famous John Wesley, founder of the Methodist Church, in the high oval set in the break arch. He died in 1791, so this could have been a commemorative dial. The clock maker, William Wain of Alfreton, is not listed. The corner decoration as well as the side pieces of the break arch appear to be gesso work, but they are all flat gold leaf, shaded to appear in relief. The main ground colour is an off white instead of the more usual Wilson pale blue, and this colour tended to be used in the later dials. The surface is softer, and more liable to fine cracking as well as dirt absorption.

The new or all Arabic style of graphics has been mentioned before and this dial (**Illustration No. 19**) is a good example of the so-called tumbling Arabic numerals. The clock maker, James Isaacs of

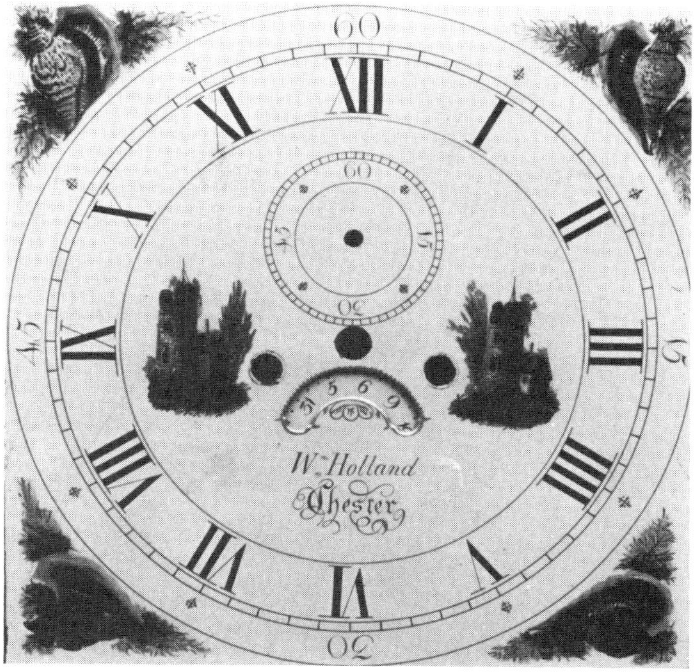

13 The latest type of Wilson dial with shells in the corners and castles in the centre. It was made for William Holland of Chester *c.* 1809. See page 35–6

Carmarthen (1805–20), must have been one of a small Jewish settlement there as the Samuel and Lazarus clock makers also lived in Carmarthen. The surface of this dial also was off-white, and it was probably made *c.* 1805.

There were regional differences in dial sizes and shapes. In the northern parts of Great Britain, roughly above a line from Bristol to Norfolk, dials tended to be larger than in the south. Thirty-hour clocks usually had smaller dials than the eight-day ones. Round dials were very popular around Sheffield and Leicester as well as in Scotland. Groves Donishorpe of Hinkley used a round dial about 1785 and this dial had a complete job ticket on the back (**Illustration No. 20**).

One of the problems for the clock maker was the cost of the case which had to be made by the cabinet maker. An arched top moon dial case could be expensive, though the same would be true for the break arch dial. An ingenious solution for those customers who wanted a square dial and also a moon was the dial with the '12 o'clock moon'. This style was popular in Cheshire, Lancashire and Yorkshire. An exceptionally fine dial of this sort (**Illustration No. 21**) was owned by Beatrix Potter and was illustrated in *The Tailor of Gloucester.* This

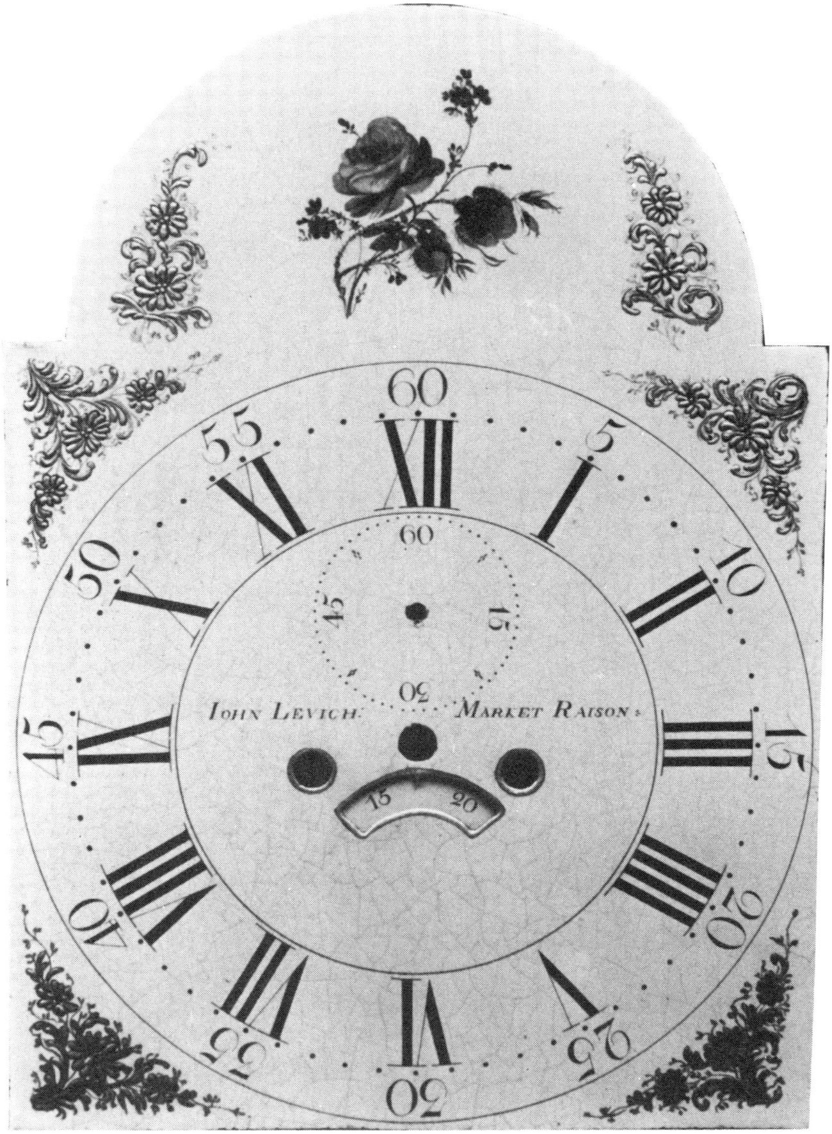

14 Early Wilson arched dial of late 1770s with four long feet and a date dial on a stub foot. It was made for John Levitch of Market Raison. See pages 36 and 37

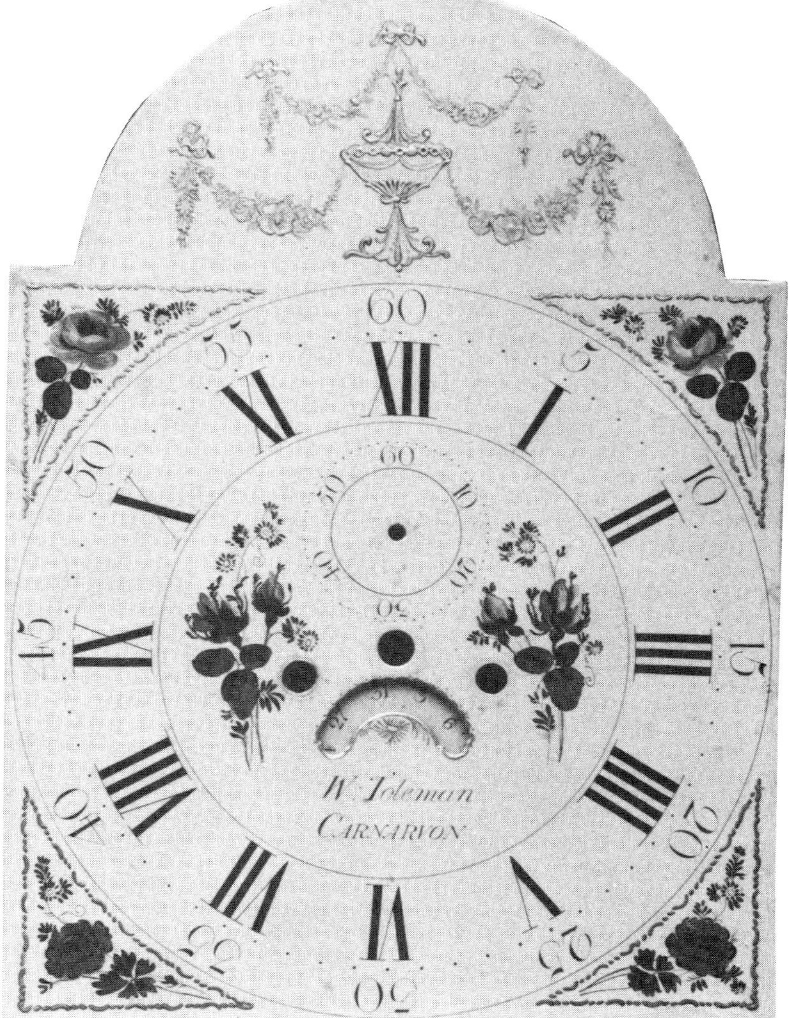

15 Elaborate Wilson decorated break arch style dial made for Toleman of Carnarvon *c.* 1790. See page 37

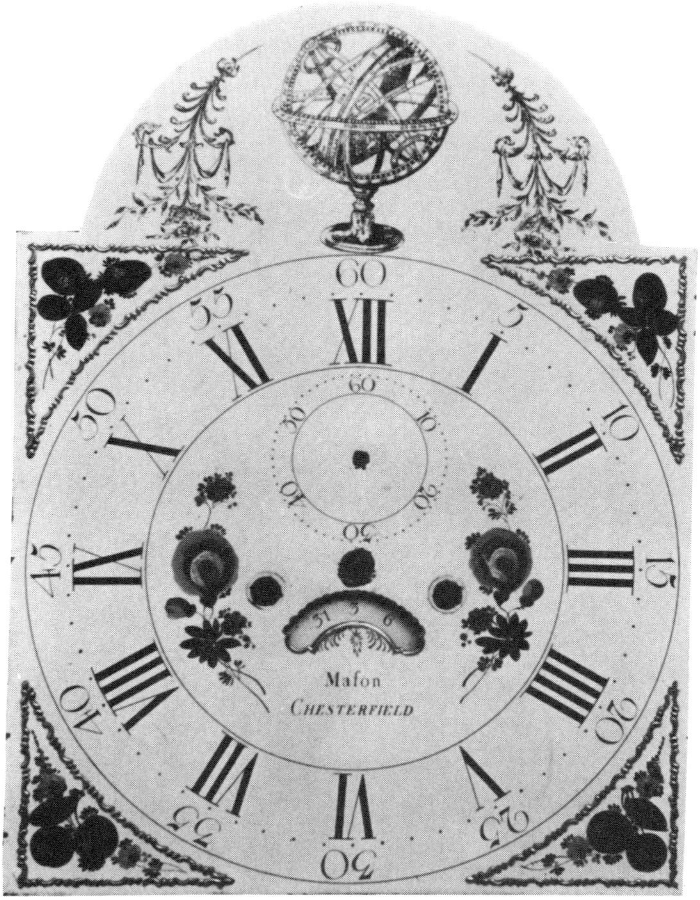

16 Unusual transfer print of armillary sphere in the break arch made by Wilson for Timothy Mason of Chesterfield *c.* 1790. See page 37

handsome clock can still be seen at Beatrix Potter's house in Sawry, administered by the National Trust, in the Lake District. The only date given for the clock maker, Thomas Barrow of Stockport, was 1789, which seems about right. The gesso raised work framing the corners is exceptionally fine. The painting of the moon dial is also of very high quality, and the placing of the moon's age numerals on the dial was rather unusual. More commonly, these numerals were put on the main dial. The 'humps' on this dial have garlands, but other dials of this type used maps, though because of the size the full map could not be put on.

The next section deals with arched top with moon dials, sometimes called 'Rolling Moons'. Except for automata, these dials were the most expensive. A great many of these arched-top, moon dials survive and a number of them were exported to the United States during the late

1700s and early 1800s, where they were very highly regarded. The earliest of this variety of dial often had gesso corners (**Illustration No. 22**) such as this one made for Henry Waring of Warrington (only date given 1776, when he married). It had the early type of date aperture and gilded winding holes. The gesso work was very fine, and the centre of the oval was painted dark blue. (An almost identical type of corner illustration is in Brian Loomes's *White Dial Clocks – The Complete Guide*, p. 27.) The moon dial had a starry sky, complete with comet, and the other side of the moon dial depicted a beautiful lakeside scene. The cast

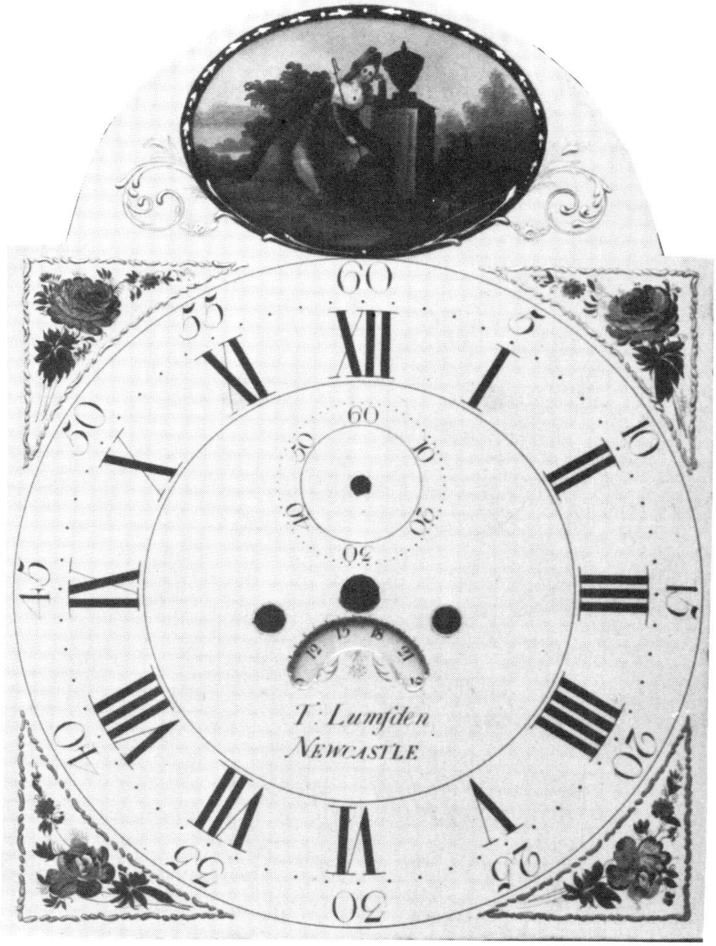

17 The lady beside the urn has appeared in other Wilson break arch dials – sometimes without the urn. This was made *c.* 1780 for T. Lumsden of Newcastle. See page 38

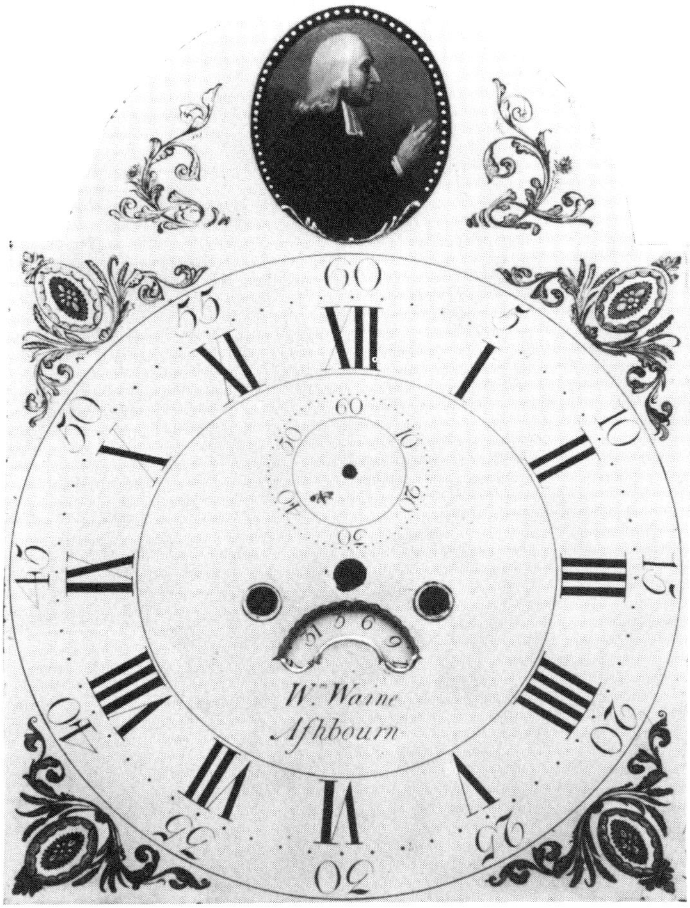

18 Unusual Wilson commemorative dial featuring John Wesley, founder of Methodism, who died in 1791. There is no real gesso, only flat gold leaf. See page 38

iron false plate had 'Jas Wilson' on it in small capitals, which seems to denote an early dial, as nearly all the Wilson false plates used large capitals. The maps are of the early type, relatively simple and cut off so as not to interfere with the corner decorations. Another almost identical dial exists with rose pink centres to the oval gesso work, but it has no false plate and four long feet.

A similar type of corner was used in a later dial *c.* 1790 by John Stanyer of Chester and Nantwich (1790–5) (**Illustration No. 23**) where the maps extended to the edge of the chapter ring and the date aperture

19 Wilson used 'new' type of all Arabic numerals in this dial made for James Isaacs of Carmarthen *c.* 1805. See pages 38–9

was of the rounded end type. The moon dial had two land scenes instead of the more usual Wilson type. Regional variations have been mentioned and one of them can be seen in this dial (**Illustration No. 24**) made for John Thomson (only date given is 1789) of Lesslie, where the tidal dial was put onto the moon dial in small Roman numerals. This use of tidal dials seemed to have been quite popular in Scotland and often there was a pointer of some sort attached to the dial. The break arch motto was not all that common. The name and town were not put on at the Wilson manufactory, but probably done locally. The maps were of a later type,

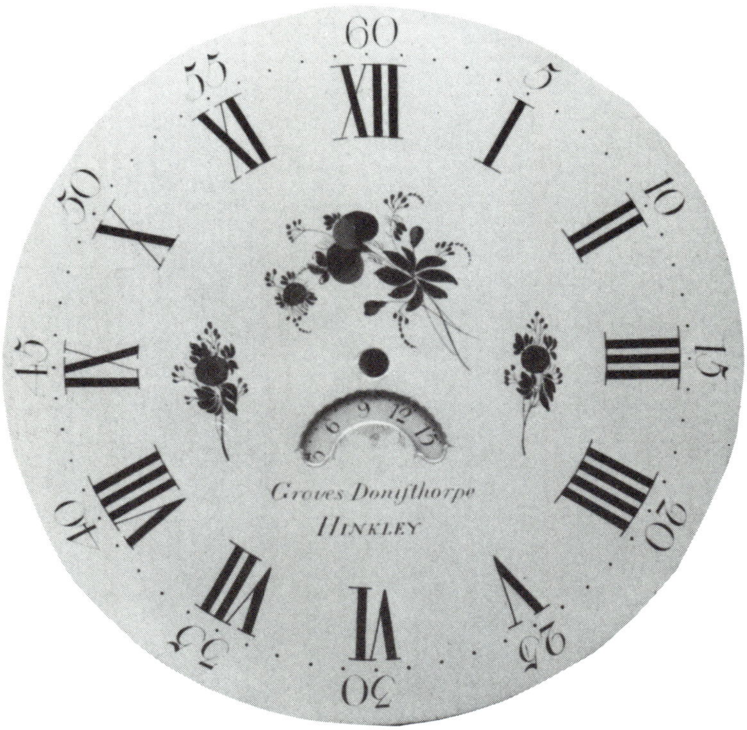

20 Round Wilson dial, popular in Sheffield and Leicester. This was made for Groves Donisthorpe of Hinkley *c.* 1785. See page 39

where North America was a bit wider and the ecliptic was curved below the equator on the map of the Americas. Another regional variation was the use of centre date numerals instead of a date aperture or date circle on the main dial. This was popular around Liverpool and also in the US state of Pennsylvania.

It is fascinating to trace the work of one particular artist. In the dials

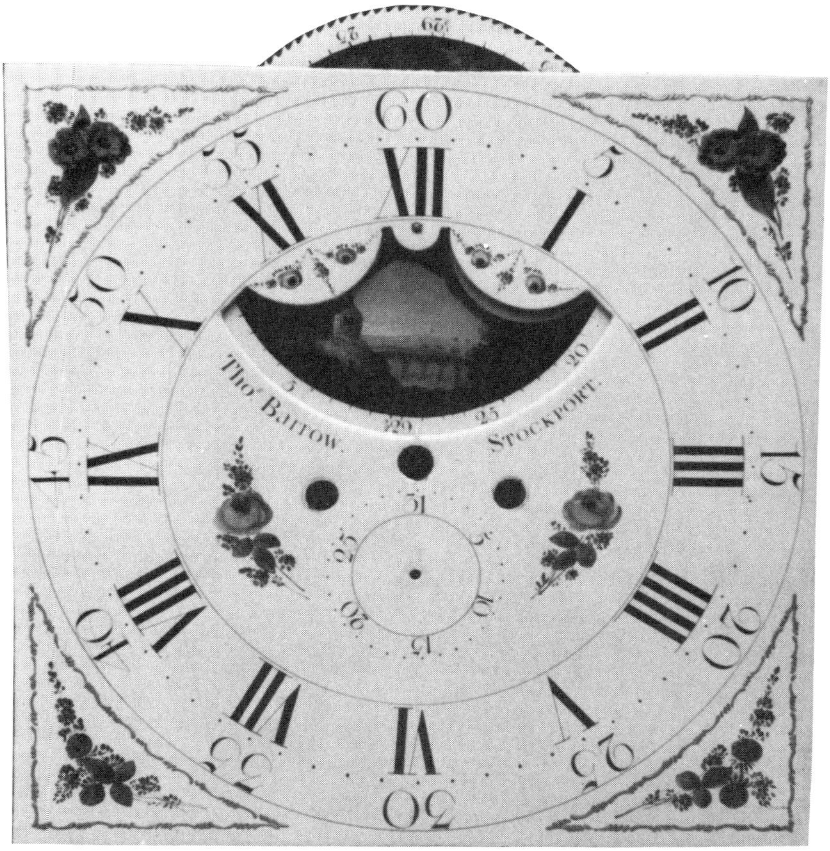

21 Beatrix Potter's own clock was used in her illustrations for *The Tailor of Gloucester*. This is a very fine Wilson twelve o'clock moon type of dial *c.* 1784, with finely detailed gesso framing the corners. See pages 39 and 42

where the four seasons and other themes are depicted in the corners, there seem to have been two main types of figure painting. In the earliest variety, the ladies had rather pale faces, similar to this dial (**Illustration No. 25**) which was made for William Jackson of Frodsham (dates given as 1795–1801) about 1790. Probably the most famous Scottish clock maker was John Smith of Pittenweem (1770–1814), who made some very costly and complicated clocks, often using Wilson dials. One of these is beautifully illustrated in Cedric Jagger's *Royal Clocks*. It is a four-dialled clock with both main dial and musical side dial by Wilson. The artist appears to have been the same as the one who painted the Jackson dial. On one of the other sides of this clock there is a moving automata with a procession of figures appearing through two doorways. The celebrated artist Alexander Nasmyth was supposed to have painted

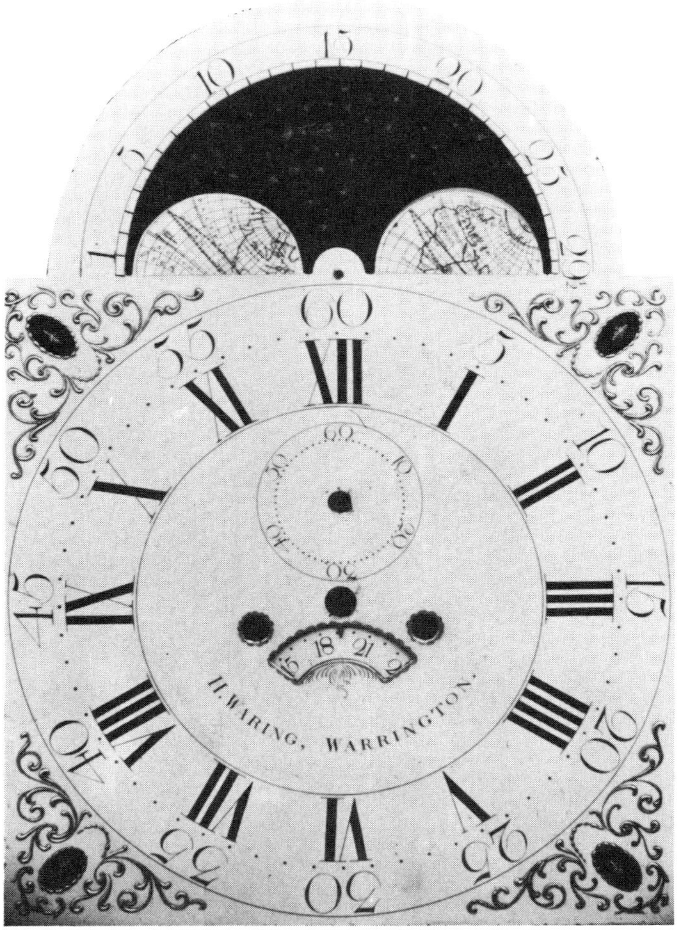

22 Early type of Wilson break arch moon dial with early cut off maps, square-ended date aperture and gilded winding holes. It was made for Henry Waring of Warrington, 1778–9. See pages 43 and 44

this. Another John Smith clock is a real giant, standing at least 9 feet tall. It can be found in the National Museum of Scotland, Edinburgh, and has the same type of pale-faced ladies, though the side panel with automata is quite different and appears to have been painted with a rather unskilled hand. This particular dial is also rather unusual in having a small vitreous enamel applied ring with the names of the tunes on it in the lower part of the main dial.

Both Smith clocks have corner paintings on the main dials depicting Justice or Law, Medicine, Trade or Exploration and Military, though the figures and backgrounds are slightly different. The second and

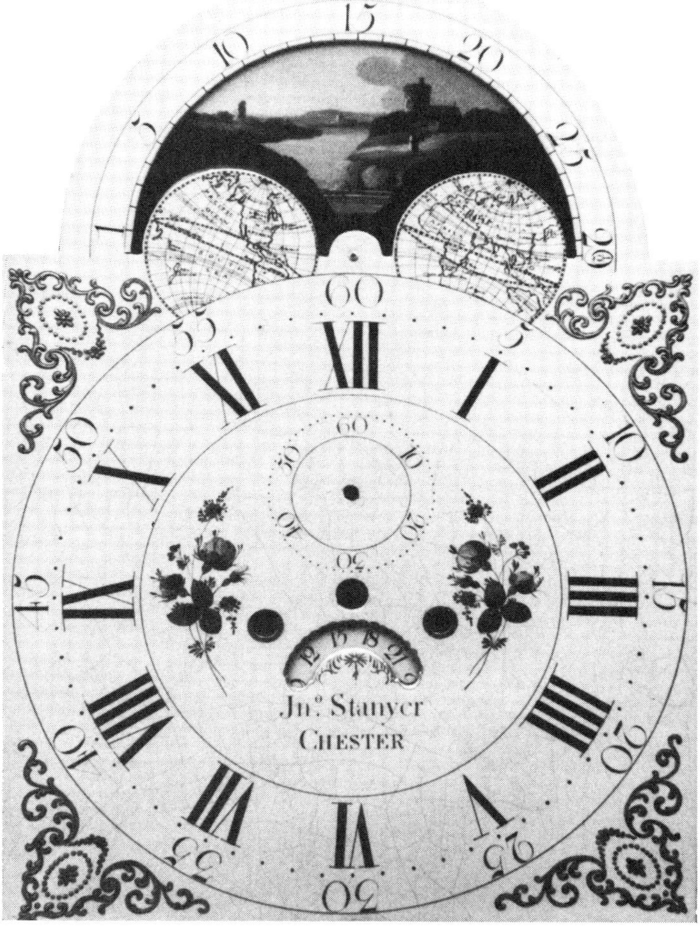

23 Late Wilson break arch moon dial with similar corners to Illustration No. 22 but featuring rounded date aperture and more complete maps. It was made *c.* 1790 for John Stanyer of Chester. See pages 44 and 46

probably later type of figure painting had a more romantic look, where the ladies' expressions were rather sweet, if not actually simpering. **Illustrations Nos. 26 & 27** show two dials where the corners are almost identical except for the lower right. The presumably earlier dial was made for William Lassell of Toxteth Park (1758–90 or 1807) and had an unusual early fully-painted centre as well as centre date numerals. The use of the centre date hand meant that there was no date aperture or small date ring to fill up the middle part of the dial so that the artist had a clear field in which to express himself. This dial was probably made

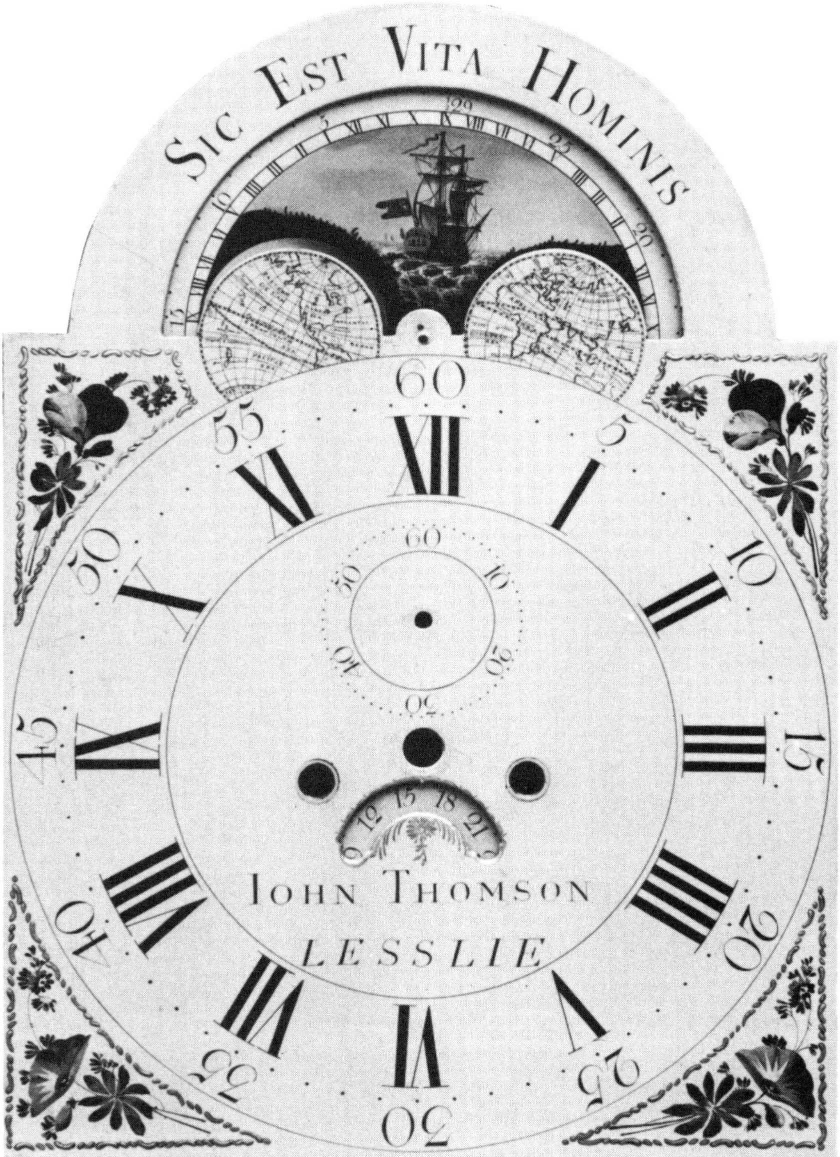

24 Scottish clock makers preferred the tidal type of moon dials so Wilson made quite a few of this sort of dial for the Scottish market. John Thomson of Lesslie (sic) had his name added locally *c.* 1789. See pages 46–47

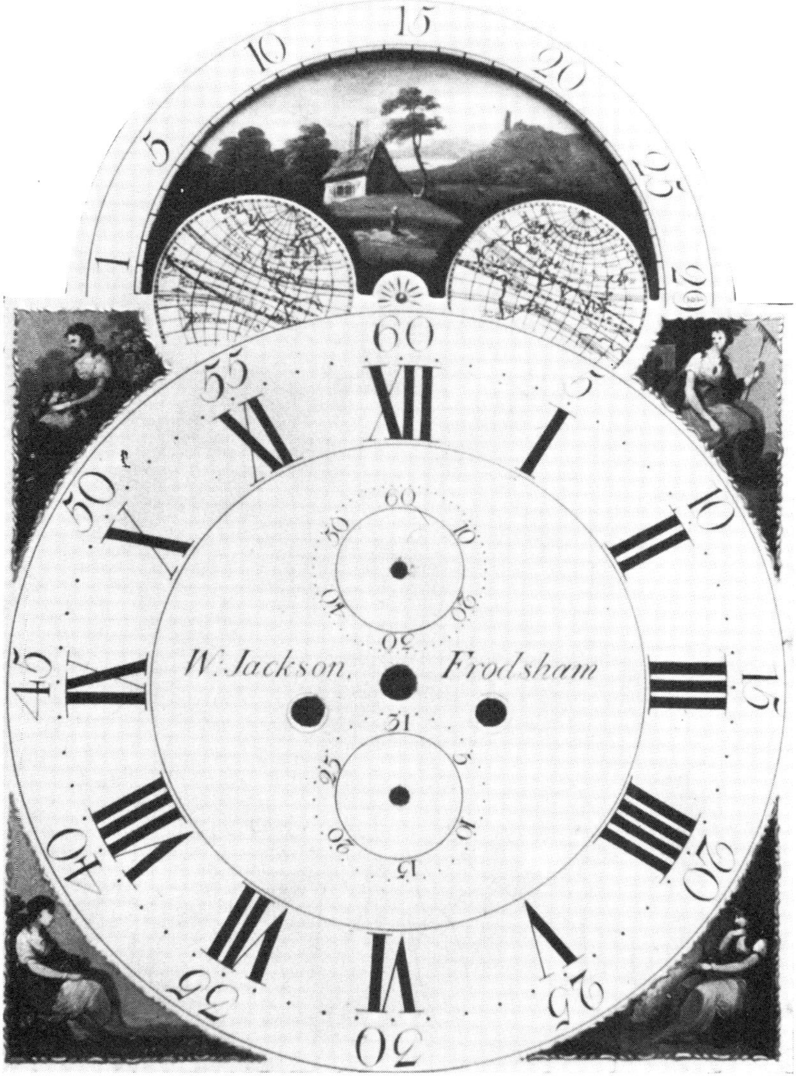

25 A Wilson four seasons break arch moon type of dial with pale-faced ladies similar to those found on a John Smith of Pittenweem dial. William Jackson of Frodsham was working about 1790. See page 47

about 1795, but there is, unfortunately, no movement or case. The second dial was made for William Waight of Birmingham (1790–1835) about 1800, as it used the 'new' style Arabic numerals with a partially dotted minute ring. This type of minute ring shows the transition from the dotted sort to the band with bars. The author has restored another dial from this period, also made for William Waight, where the four seasons paintings are quite different though obviously by the same artist. This shows that there must have been a considerable amount of latitude given to the individual artist, who never painted two dials exactly the

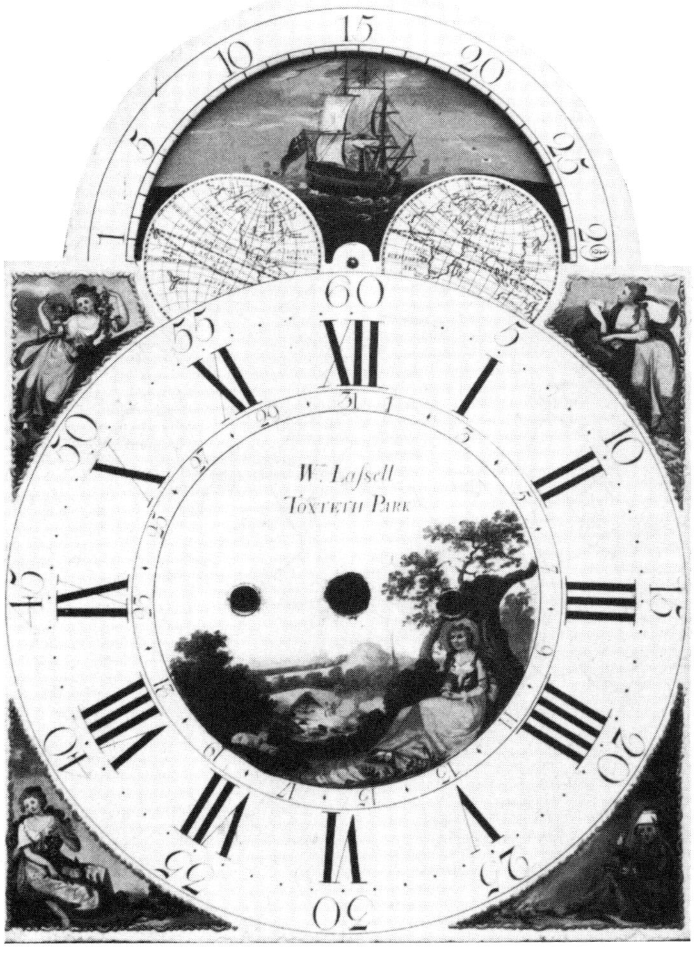

26 A Wilson four seasons break arch moon dial with rare centre painting made for William Lassell of Toxteth Park, Liverpool *c.* 1795. See pages 49 and 52

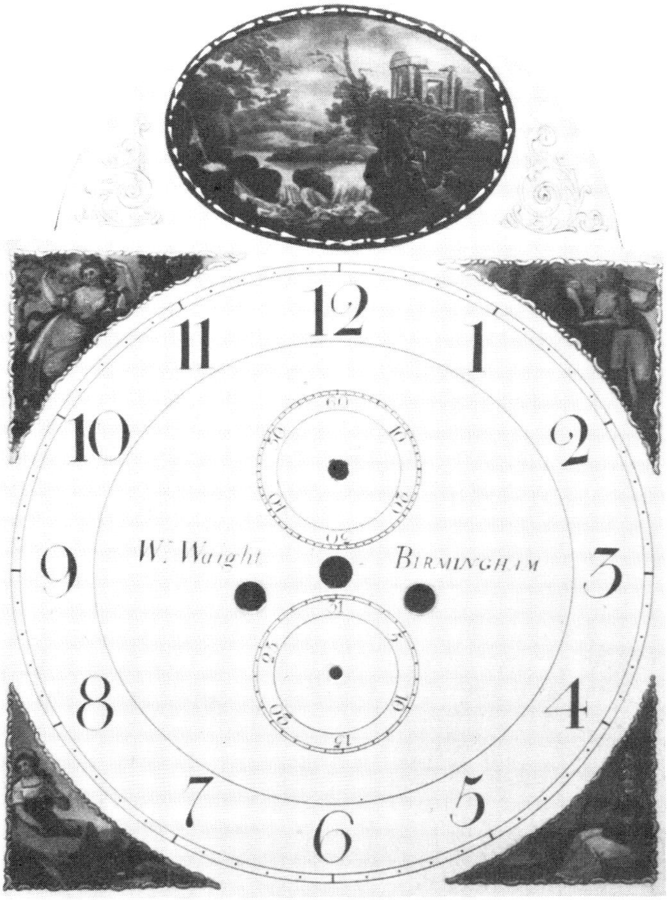

27 A Wilson dial where the four seasons corners were painted by the same artist as No. 26. The numerals are Arabic and the minutes are marked with dots and bars in a minute band. William Waight of Birmingham ordered this dial about 1800. See pages 52 and 53

same. Though the Lassell and Waight corners are very similar, the colours of the ladies' dresses are quite different.

Quite a few Wilson dials featured a green background to the gesso-framed corner paintings of fruit and flowers from 1790 to 1800. This green colour has often started to degrade and can look very blotchy.

In general, all of the style changes noted in the square dial were repeated in the break arch moon type of dial. The closer one gets to the end of Wilson's production run that finished in 1809, the more frequent the use of fan corners, with or without vignettes, baskets of flowers instead of single flowers, or even roses done entirely in gold leaf such as seen in the dial (**Illustration No. 28**) made for Henry Wardlaw of Liverpool (1805–29). This dial *c.* 1806 used the flared Arabic numerals and the earlier flowers in the centre have been replaced by swans which give a heavier effect. The maps were of the late Wilson type. The cast iron false plates from about 1800 often had Wilson on both sides, possibly to make sure the name could be seen even when the feet were put on to the wrong or blank side of the false plate.

Oval Wilson dials were certainly not common because the special hood shape was expensive, but they definitely were made. They can be

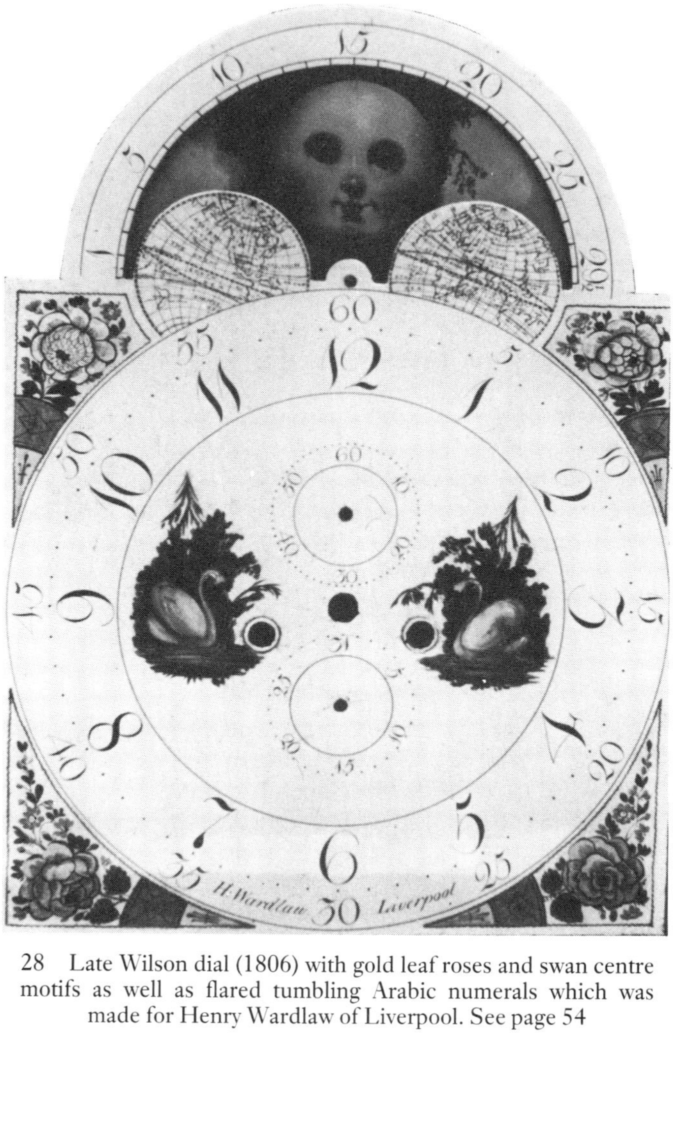

28 Late Wilson dial (1806) with gold leaf roses and swan centre motifs as well as flared tumbling Arabic numerals which was made for Henry Wardlaw of Liverpool. See page 54

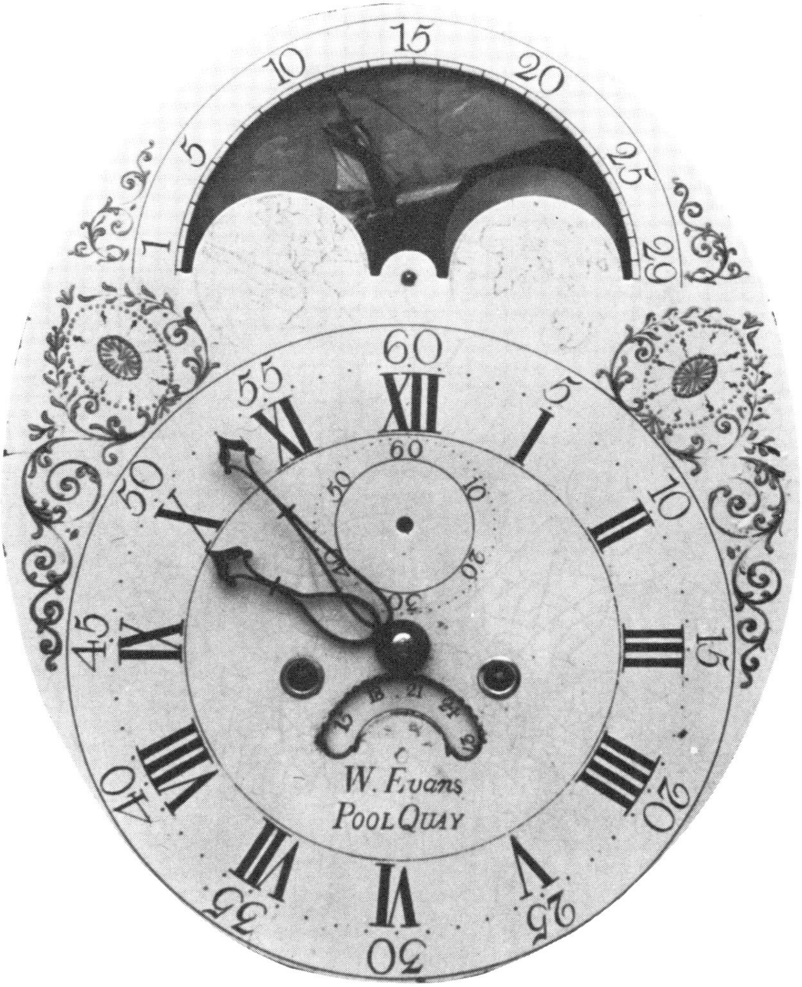

29 Rare oval Wilson dial made for William Evans of Pool Quay *c.* 1790. This dial appears to be partially unrestored as most of the maps are missing. See page 56

divided into several categories. Some had moon dials in the top part, like the one made for James Lomax of Blackburn illustrated on p. 31 of Brian Loomes's *White Dial Clocks*. Others had an oval scene at the top on a vertical axis, and this scene often contained a female figure with various swags, rosettes or urns in gesso surrounding it. Another variation could have been a large scene painted all over the top third of the dial. Although as yet the author has not restored an oval dial, one most attractive dial of this type was made available for photography. **Illustration No. 29** shows elaborate gesso work in the corners of a dial made about 1790 for William Evans of Pool Quay. This large, rectangular wall clock was of an unusual type, though most oval Wilson dials were made for long case clocks.

Samuel Deacon of Barton in the Beans (1771–1816) finished what some would call his masterpiece on 9 October 1790. It was an 8-day musical clock with a 19″ × 15″ oval dial. In the arch were automata of eight musicians as well as a boy and girl. This shape of oval was rather wide, and had gesso scrollwork at the sides of the chapter ring. There exists a letter from Samuel Deacon to James Wilson about this dial where he states that he wants 'the face to be painted well and soon', plus a complaint that in the past he had lost money and had 'great vexation' due to the slowness of the work. This very complicated Wilson dial cost £15 15s, which was expensive at that period. Some not very clear photographs of this clock dial can be found in P.A. Hewitts' article on 'The Deacon Family of Leicestershire Clock Makers, Part II' *Antiquarian Horology*, 1986, p. 350.

Wilson dials with automata are comparatively rare because, outside of the 'special orders' such as the Deacon oval dial, they were the most expensive of all dials. Few seem to have survived with all of their moving pieces intact, but one that does exist is of the 'Adam and Eve with arms to move' type (**Colour Illustration (iii)**) made for Thomas Mear of Dursley (1789–93). The artist's mark was on the back ∵ and the detailed carnations in the top corners seem to have been his speciality. This dial contains a well-painted fly which covers an imperfection or chip in the dial. Quite a few of these 'fly dials' exist, and in every case the fly covers a defect. As far as is known, only Wilson used this device.

Bracket clock dials roughly followed the same style of both graphics and decoration as their larger relatives. The author has restored a very fine dial of about 1780 that was in all ways, except for the standard-sized winding holes, a miniature break arch moon type, complete with gesso scroll corners, maps and a fine moon dial. The flat gold leaf work in the corners was also used in bracket clock dials made closer to 1800.

This concludes the rather exhaustive survey of James Wilson dials. The author has attempted to provide a framework so that different types of dials can be classified. Each dial maker had his own characteristic style, but it is not always possible to definitely attribute an unmarked dial to one particular maker. However, the more that is known about each maker the better the possibility of correct attribution.

THOMAS HADLEY OSBORNE DIALS AND ALL THOSE MARKED OSBORNE 1777–1813

James Hadley Osborne carried on working after the partnership with Wilson was dissolved in 1777, but he appears to have died in 1779. In various forms, the firm carried on. First came the widow Ann Osborne (1780–6), then Ann and son (James) (1787–90); next Ann and Co. (1791–6), later Ann and James (1797–1807) and finally James alone (1808–13). Although not producing nearly as many dials as Wilson, the Osborne firm made quite a number, all of which were quality products. Many of these Osborne dials were exported to the United States.

Although Osborne dials seem similar to Wilson ones, there are some subtle differences. The gesso borderwork is usually of a different pattern to that used by Wilson, though sometimes the dot dot dash was used. His unique dial (**Illustration No. 30**) has typical Osborne gesso work in the corners. It was made for one of the Fleetwoods of London *c.* 1780, and it is rare that a London clock maker wanted a painted dial. The whole dial is a special one with very small minute numerals, a 'strike-silent' and the profiles of Plato and Socrates in the break arch. Interestingly enough, these same profiles done in transfer printing turned up on the break arch of another Osborne dial which was made slightly later.

The fully painted arch scene was an Osborne speciality, like this one (**Illustration No. 31**) made for Thomas Griffis of Birmingham *c.* 1782. The winding holes are gilded, but done with slightly more elaboration than those of the Fleetwood dial. The hunting scene was very well painted, and the gilded gesso surrounding it is of the usual Osborne pattern.

Another fully painted break arch scene with the painting done right up to the edge of the break arch was made for one of the Winstanleys of Holywell. The winding holes were decorated in crimson lake with swags, and a floral decoration below the date aperture. This treatment of the winding holes is most unusual and has not been seen before.

Some clock makers remained resolutely old fashioned and continued to use the square date aperture with a silvered dark ring held on by rollers that had been used with the brass dial. In the case of the painted dial, this meant a special job because the attachments to hold the rollers had to be put on before the dial was painted. James Robertson of Dundee (**Illustration No. 32**) was one of those who held on to the old ways. This dial *c.* 1785 had very elaborate gesso work, but the name and town were not put on at the manufactory, so it probably was done locally. On the subject of signatures, it is commonly believed that the clock maker himself signed the dial. In very rare instances this may have been true, but generally the name and town were put on at the time of

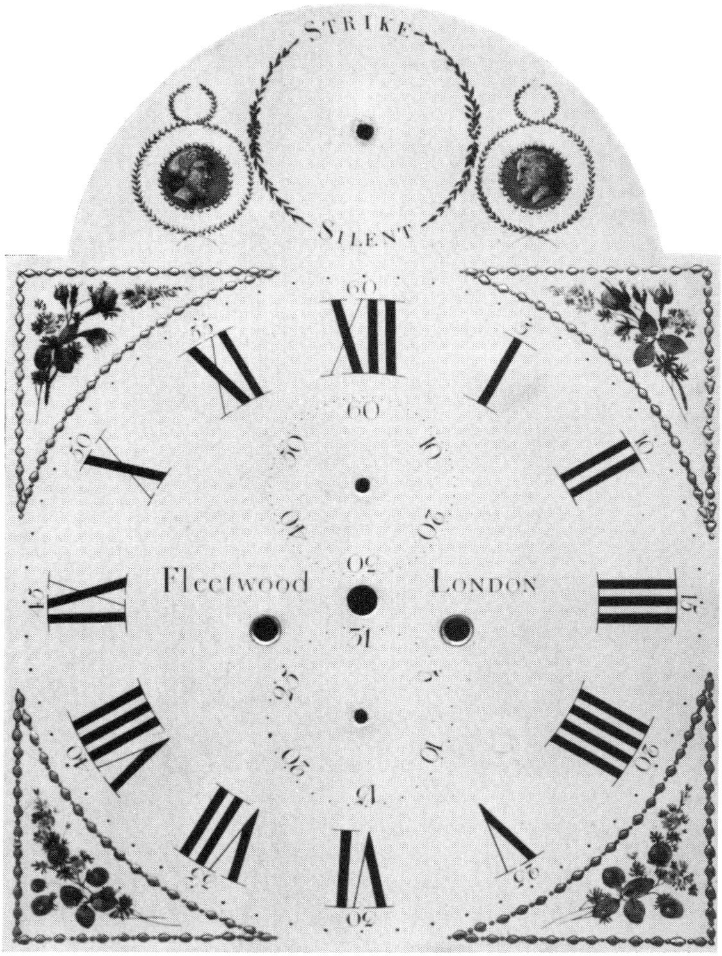

30 Unusual Osborne dial made for Fleetwood of London *c.* 1780 with profiles of Plato and Socrates in the break arch and tiny minute numerals. Very few London makers used painted dials. See page 57

manufacture. Some dials were sent out blank without any name, but if a name was put on afterwards it was probably done by someone used to fine lettering such as a lawyer's clerk. Those names that have been put on by an unskilled hand are easy to spot.

Another Osborne speciality was the dial decorated entirely with transfer prints (**Illustration No. 33**). This particular dial made for James Lomax of Blackburn (1749–1814) arrived from Holland with an attempt at spandrel type gesso corners done in glue and gold paint which, when removed, revealed traces of the transfer printing that were

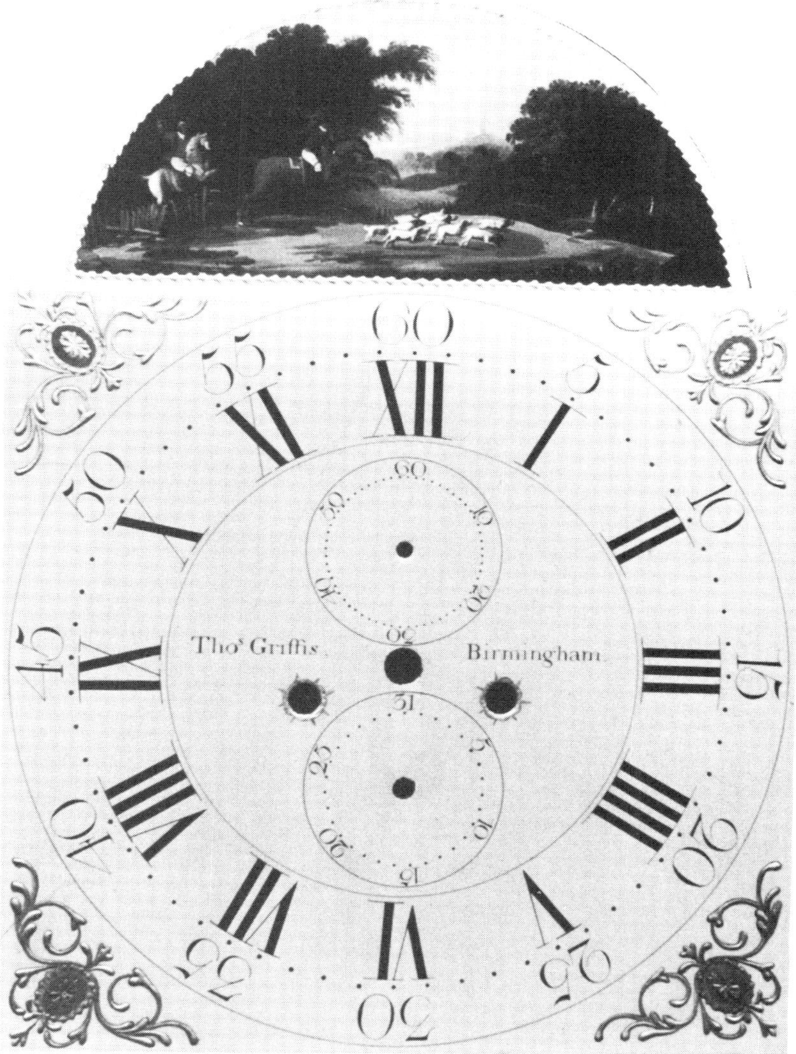

31 Beautifully painted break arch scene made by Osborne for Thomas Griffis of Birmingham *c.* 1782. The same artist painted at least one moon dial. See page 57

only visible under UV light. There were quite a few dials done with this method of decoration but, because of the nature of the ink used, many of these transfer print decorated dials have not worn well. Two distinct types of corner have been noted, and various designs were used in the

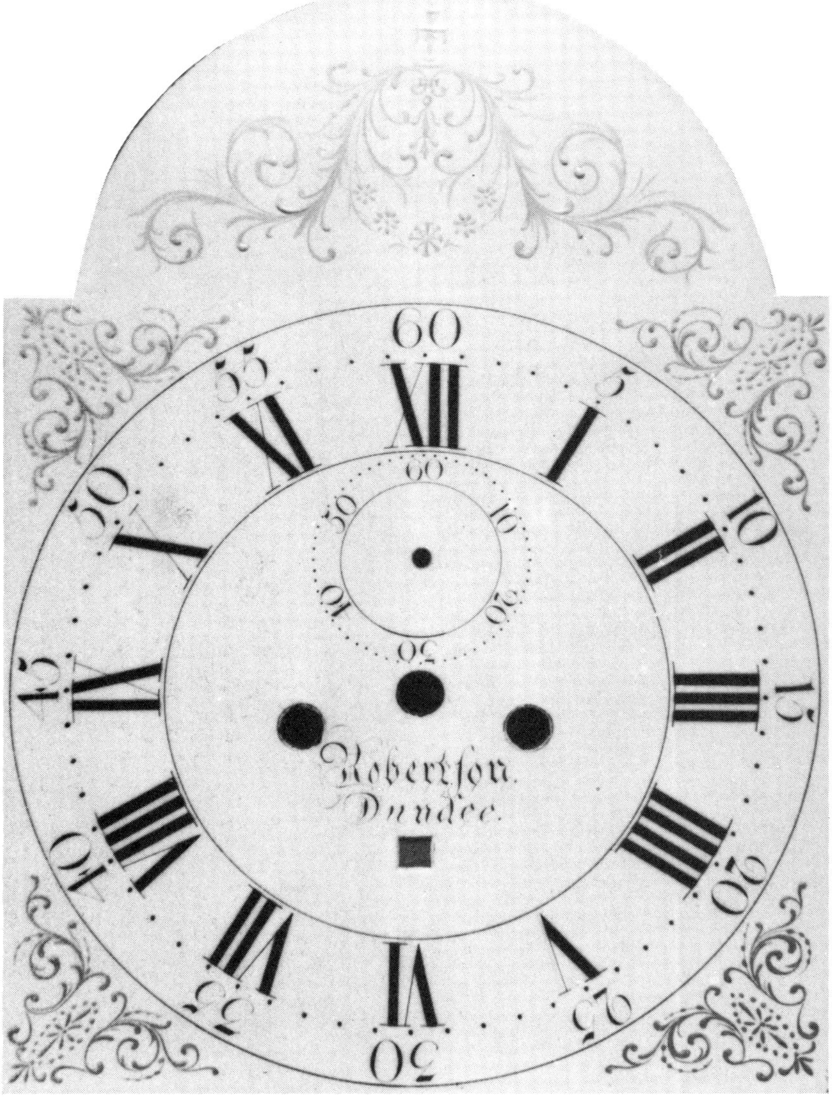

32 Old-fashioned square date aperture with silvered ring on rollers made by Osborne for James Robertson of Dundee *c.* 1785. The gilded gesso work is very elaborate and the name was added locally. See page 57

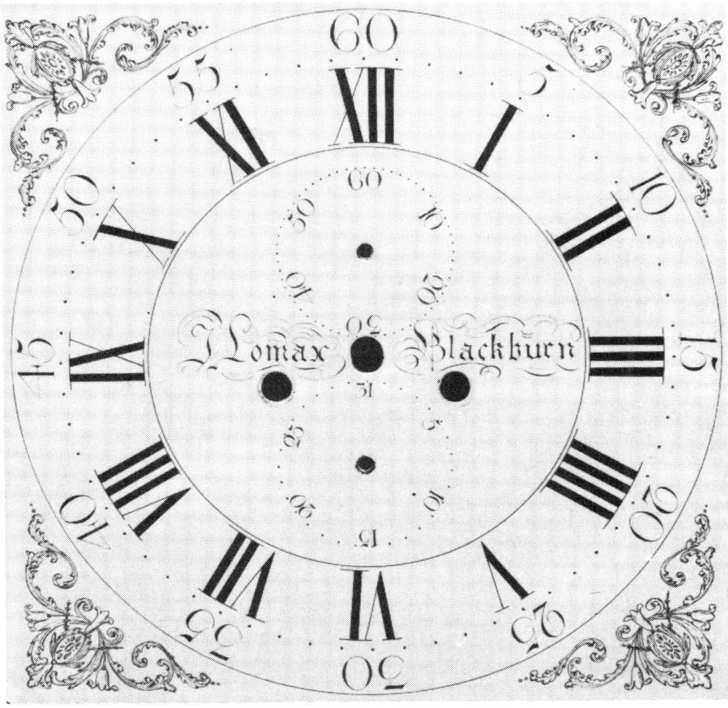

33 Osborne transfer print decoration used on a dial for James Lomax of Blackburn *c.* 1790. See pages 58 and 59

break arch. Father Time was one; another had a crowing cock in the centre with profiles of Plato and Socrates on either side, and a third type had heavy swags, ribbons and a 'husk' oval surrounding the clock maker's name. Another variety had the clock maker's name and town in a ribbon decorated with fanciful acanthus leaves. Possibly there existed different motifs which may come to light in the future, and it could be that other manufacturers, not just Osborne, made these transfer print decorated dials, because the process was universally used for putting on the maps of the world in a break arch moon type of dial as well as some of the 12 o'clock moon variety.

Moon dials gave the artist a chance for individual expression. The faces on the moons of Wilson dials were uniformly sweet in expression and the colours of the land and seascapes usually rather subdued. In contrast, many Osborne moon dials had quite brilliant colours for the land and seascapes, as well as moons that were decidedly different. Some Osborne moon faces had round eyes and in others the eyes were quite protuberant. One of the moon dial artists seems to have been rather fond of scenes depicting naval death and destruction.

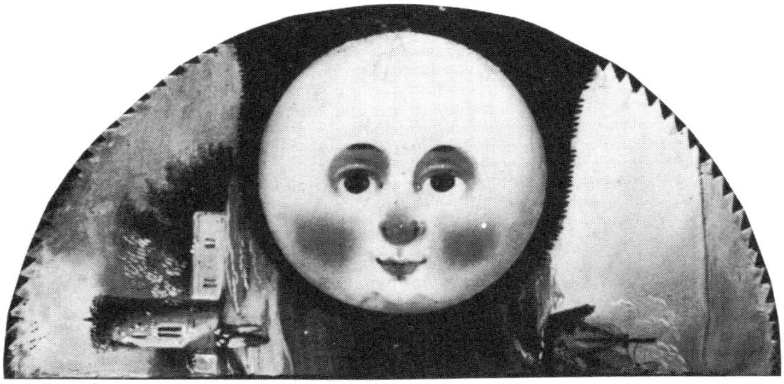

34 The economical clock maker who sawed this Osborne moon dial in half was John Stanyer *c.* 1785. Note the remains of a battle at sea on the right. See page 62

One such moon dial (**Illustration No. 34**) had been economically sawn in half by the clock maker, John Stanyer of Chester, in about 1785 because the customer presumably did not want the expense of moon works of which there was absolutely no sign in the movement. It is likely that the rest of the battle was used on another clock. Yet another moon dial, probably from the same artist (**Illustration No. 35**), depicted a night time fire at sea and was made for John Wignall of Ormskirk, also *c.* 1785. Possibly a different artist produced this amusing bit of humour (**Illustration No. 36**) where the landlady of the Swan Inn is obviously handing out a long bill. The starry sky on this moon dial was usually seen only on early (pre-1800) moon dials. It was made for one of the Barnish family of clock makers in about 1790. One can assume that the same artist painted another moon dial with identical pop eyes, but the land and seascapes were quite conventional though the colours used were comparatively bright. It could be considered that the painting of the stern landlady was a special order. Another example of this artist's work was a rather striking one (**Illustration No. 37**) featuring a trumpeting angel on a bright blue background. It was made for the well known clock maker Watkin Owen of Llanrwst *c.* 1790. The landscape was similar to other moon dials with conventional scenes, and the method of painting the tree foliage seems to be distinct to this particular artist.

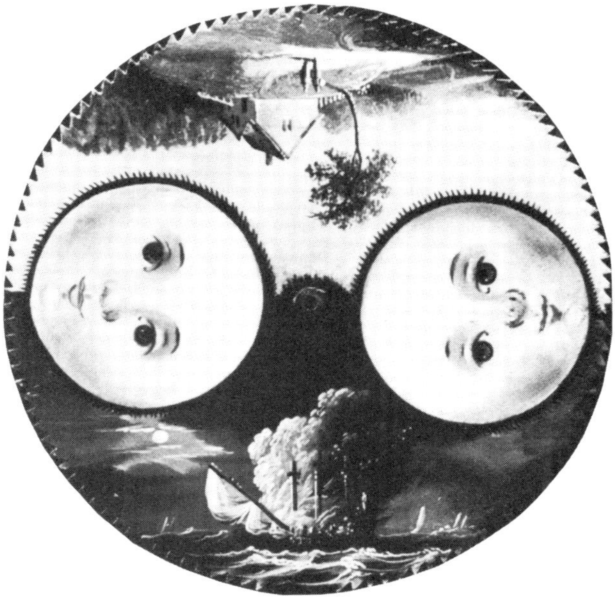

35 An Osborne moon dial painted by the same artist as No. 34 featured more destruction at sea. It was made for John Wignall of Ormskirk *c.* 1785. See page 62

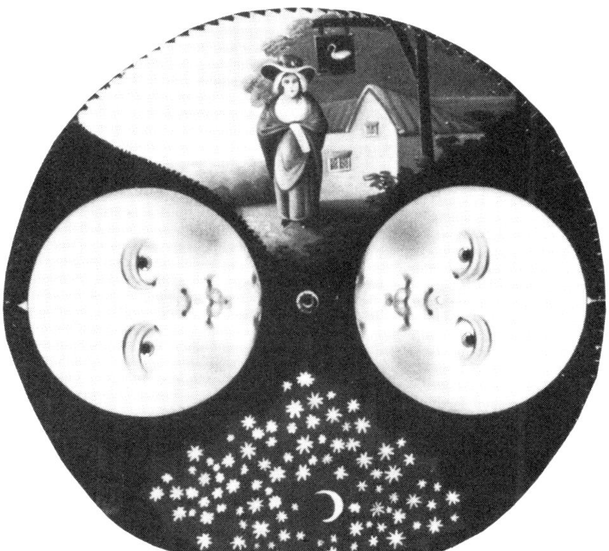

36 Humour on an Osborne moon dial made for Barnish of Rochdale *c.* 1790. The landlady of the Swan Inn is presenting a long bill. See page 62

The Osborne dials went through the same stylistic progression as those made by Wilson, though the gesso spandrel-like corners continued to be used rather later than Wilson's. There was one feature almost exclusive to Osborne break arches, and that was a bird or pair of birds in a nest. Also, when birds were used in the dial centre they were usually of the long-necked, long-tailed variety. The roses painted in the dial corners tended to have a lot of pale blue petals on an otherwise pink flower. Not many four seasons corners seem to have come from the Osbornes, but one unusual one featured boys instead of girls. The map also changed over the years, but there do not appear to have been as many varieties as Wilson used. One rare instance of maps being put on upside down happened on a 12 o'clock moon dial 30-hour clock by John Wignall of Ormskirk (1770–90). The false plates used by the Osbornes varied from 'Osborne' in flowing script to 'Osborne' in straight capitals or 'Osborne's Manufactory Birmingham', with one unusual one where the script 'Osborne' was surrounded by swags and husk decoration.

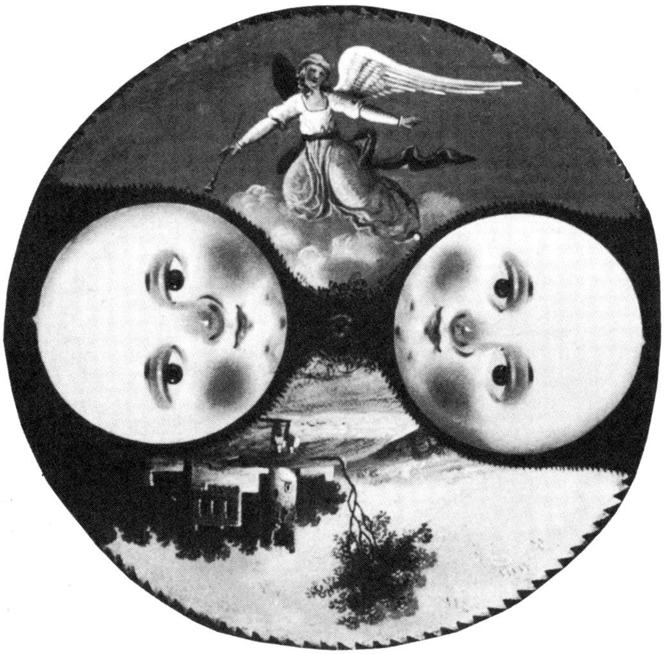

37 The same artist as No. 36 painted the angel with a long horn for Watkin Owen of Llanrwst *c.* 1790. See page 62

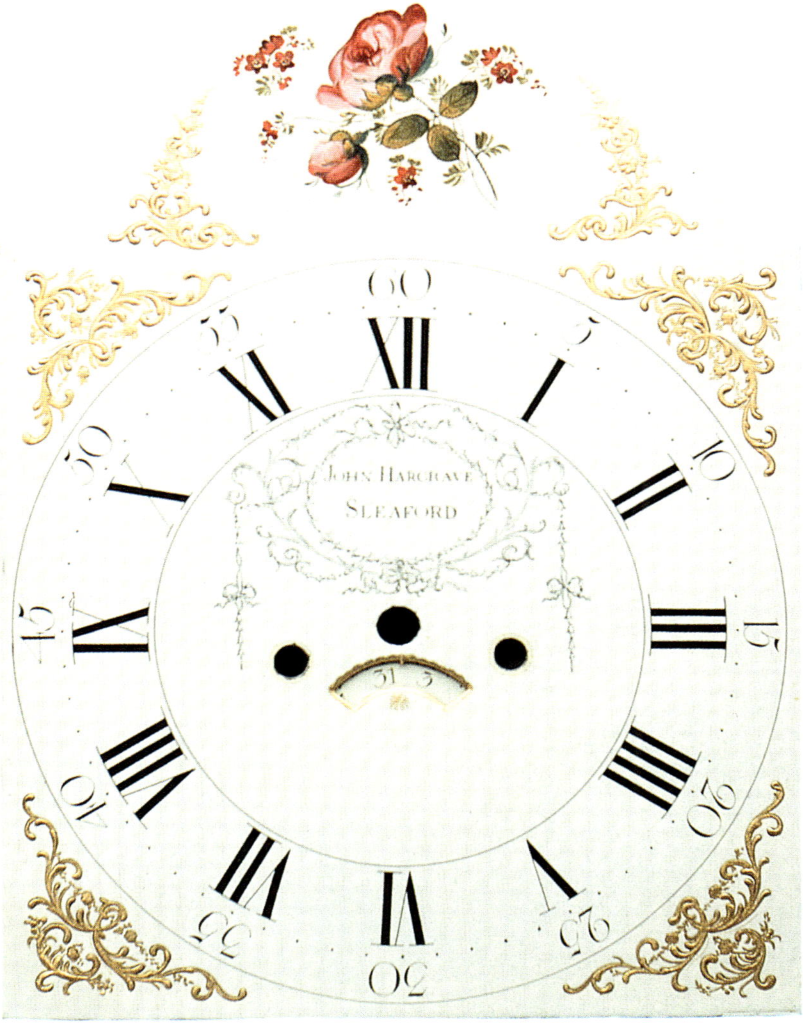

(i) John Hargrave of Sleaford, near Grantham. Wilson of Birmingham dial of late 1770s–early 1780s. Early style square-ended date aperture, early type of gesso corner decorations and first known use of transfer printed centre. This dial was originally made for a 30-hour clock, without winding holes, but Hargrave made it into an 8-day one. He even made his own crude false plate cut out of wrought iron

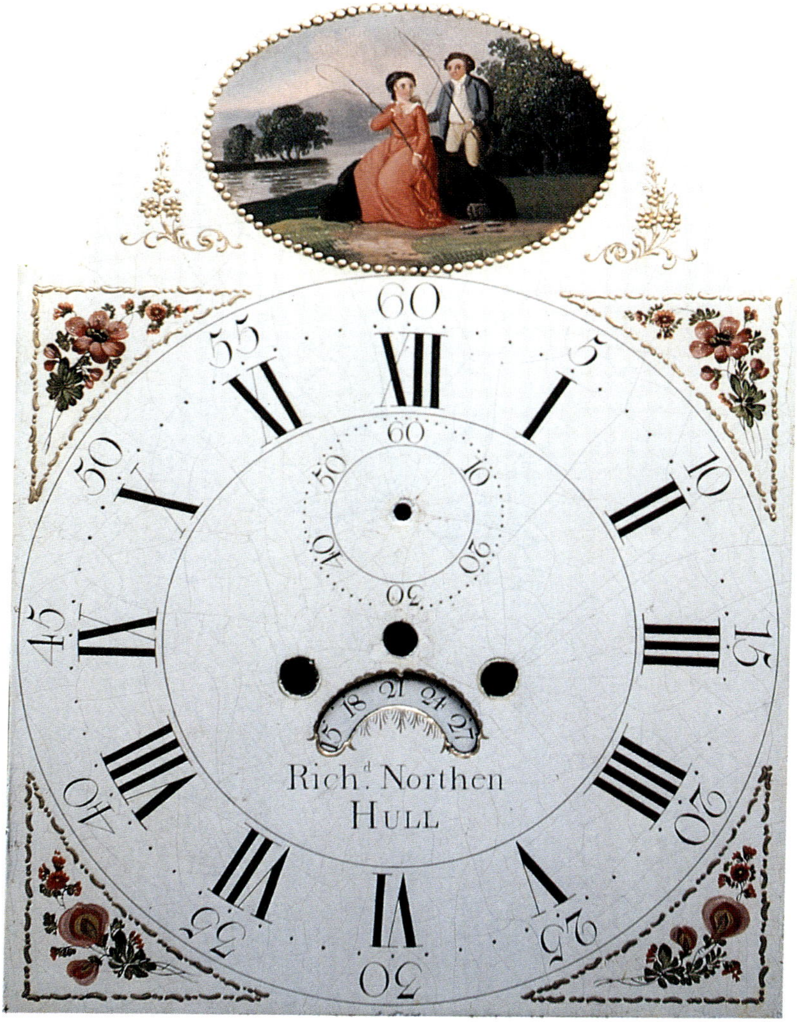

(ii) Richard Northen of Hull. Wilson of Birmingham dial of late 1780, using standard round-ended date aperture and featuring a low oval in the break arch. This depicts a couple fishing. The floral gesso decoration either side of the oval is quite restrained. The flowers in the corners, vetch and anemone, are typical of Wilson dials

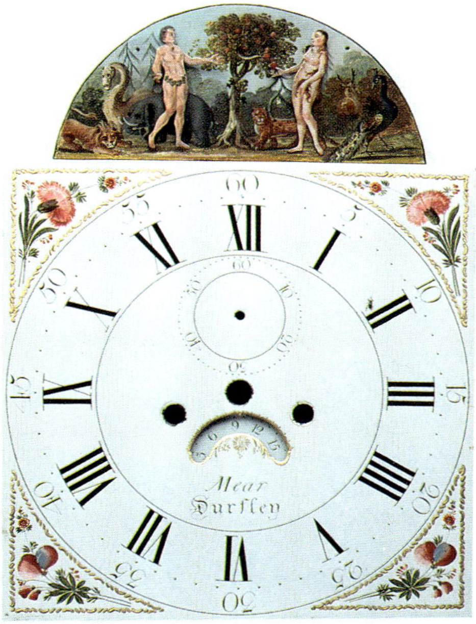

(iii) Thomas Mear of Dursley, near Nailsworth, 1789–93. Wilson of Birmingham automata dial *c.* 1790 with a fine background scene representing the Garden of Eden in the break arch. This and the main part of the dial are in one piece, and the figures of Adam and Eve with moving arms stand proud. There is a fly painted near two o'clock which conceals a defect. The dial maker's mark •: is on the reverse of the dial

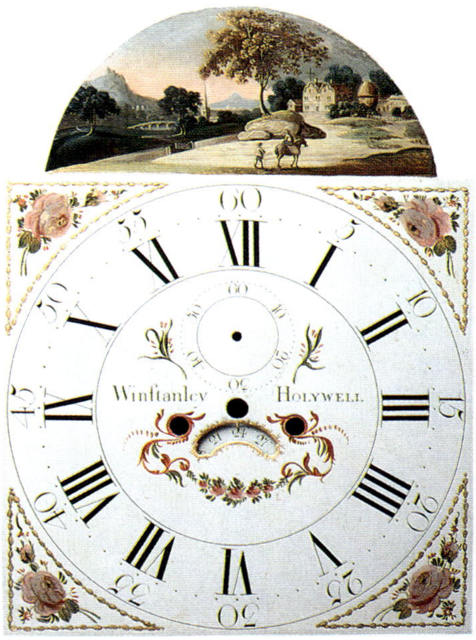

(iv) Winstanley of Holywell, Clwyd. Many Winstanleys worked in Holywell so it is hard to determine exactly which one ordered this dial, but it is an Osborne of Birmingham dial *c.* 1780–90. It is very unusual in the decorative treatment of the winding holes and the garland below the date aperture. The scene in the break arch was painted by an artist who also did moon dials. The gilded raised gesso framing the corners is of a typical Osborne pattern

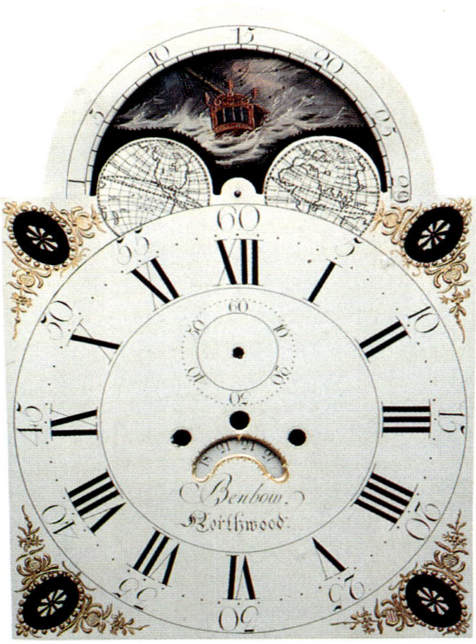

(*v*) Benbow of Northwood near Wem, Shropshire. This could have been John (*b*. 1696). It is an Osborne of Birmingham dial 1785–90 with formal gesso corners where the inside of the oval is painted dark blue. Others of this type had rose coloured ovals. There is a dismasted ship on the moon dial

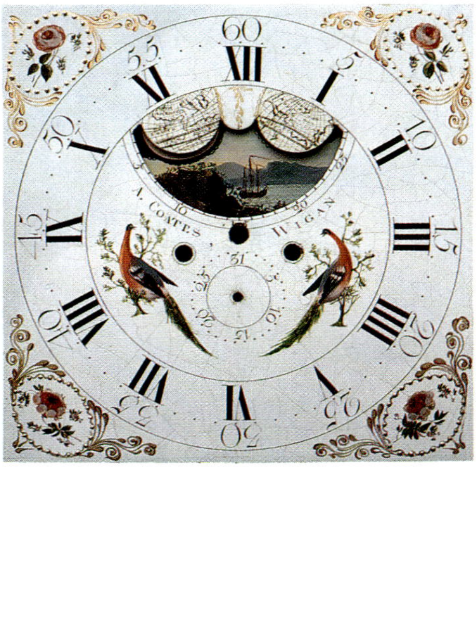

(*vi*) Archibald Coates of Wigan 1767–97. Osborne of Birmingham made this twelve o'clock moon type of dial 1785–95. It has the typical Osborne long-necked birds in the centre and some blue petals on the roses. There is both flat and raised gold leaf work in the corners and an unusual gold leaf decoration between the two maps. Only part of the transfer maps was used because of the small size of the hemispheres

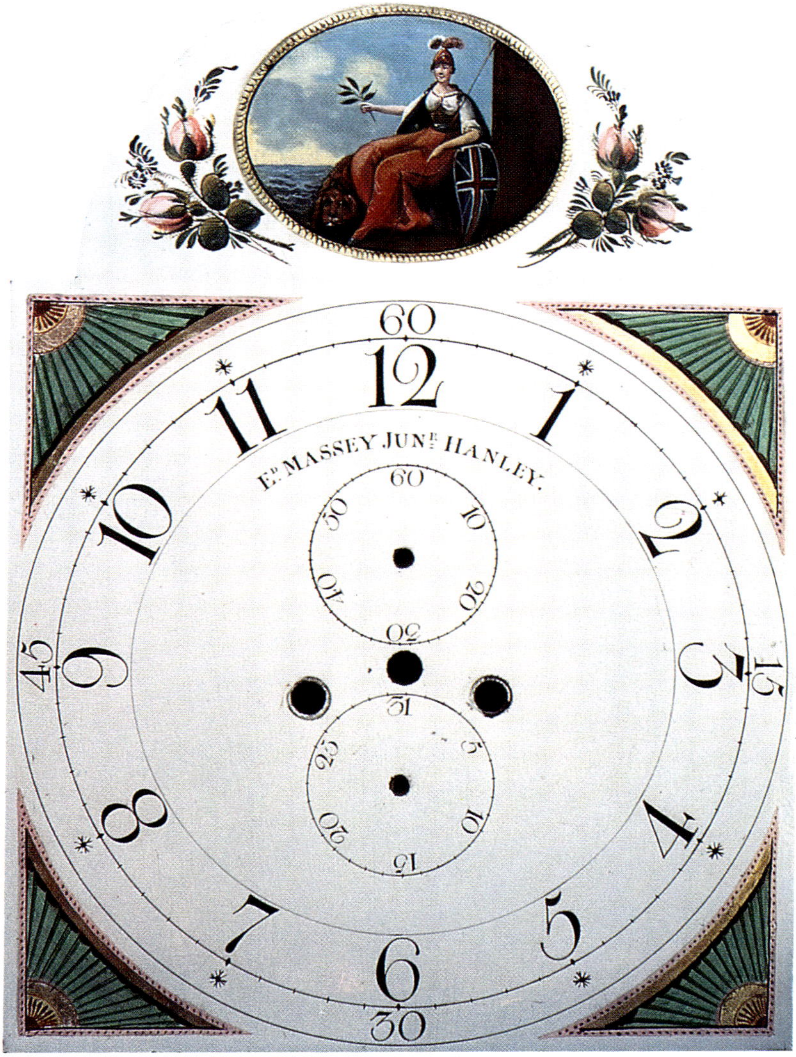

(vii) Edward Massey of Hanley near Stoke-on-Trent, 1772–1852. A Byrne of Birmingham dial with a dated movement of 1806 which has a quarter ting-tang. Fan corners such as these became popular after about 1800 when the 'new' all Arabic numerals were used. The figure of Britannia in the break arch has the typical heavy thighs found on many Byrne dials

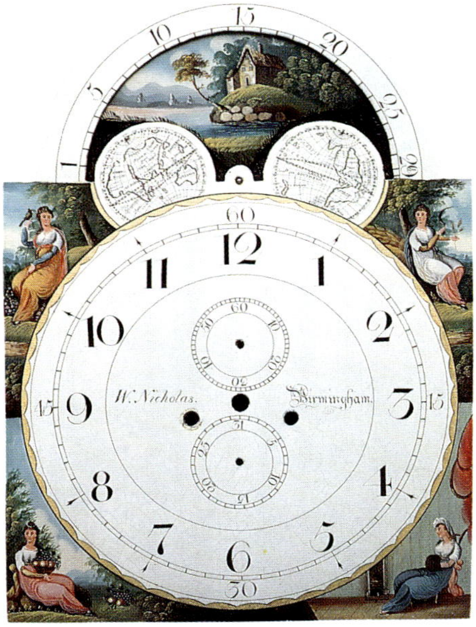

(viii) William Nicholas of Birmingham. William, Caleb and Joshua Nicholas, 1815–20 made movements as well as dials. This particular dial has an unusual scalloped gold border to the chapter ring as well as very peculiar maps on the hemispheres. These show an early use of outside numerals for the latitudes. The ladies in the corners represent the four seasons

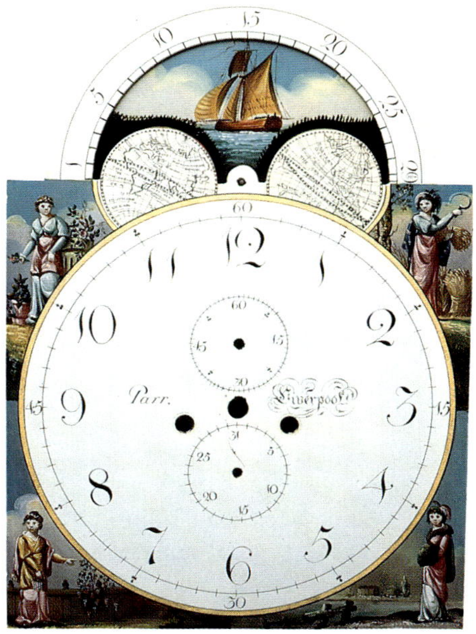

(ix) Parr of Liverpool, possibly John, 1814–34. This is a Felton of Birmingham dial, 1815–20 showing very individualistic painting in brilliant colours of the four seasons ladies who have very pale faces. The maps of the world have the Americas on the right-hand side

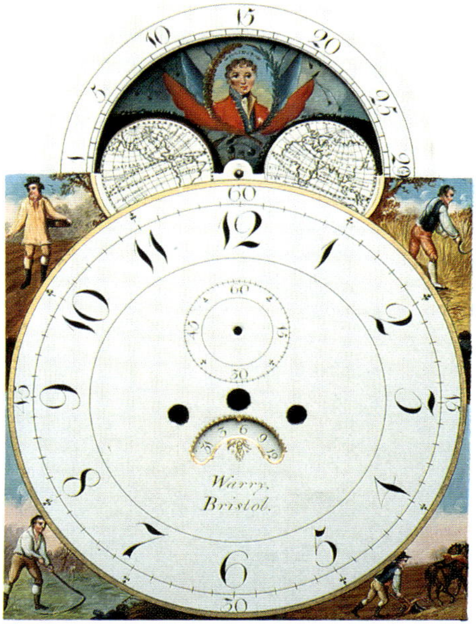

(x) Warry of Bristol, probably John senior, 1775–1830. Here is another Felton dial possibly from the same era as the Parr dials, but employing unusual male figures in the four seasons. There is an early portrait of Wellington on the moon dial which is unusual. The Americas are on the left-hand hemisphere and the ecliptic is above the equator instead of below it as in the Parr dial. This means that different transfers must have been used as the two dials differ in size

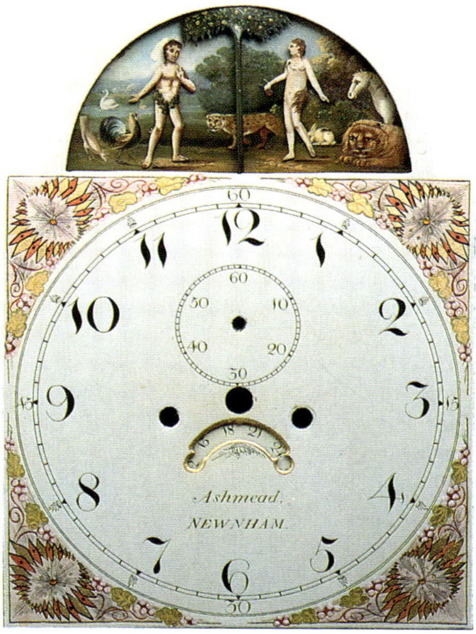

(xi) Ashmead of Newnham, Gloucestershire, is an unknown clock maker. This is a Walker and Hughes of Birmingham automata dial *c.* 1813 where the background of the Garden of Eden is a separate recessed piece and only Eve's arm moves. The corners are definitely Walker work with their repetitive diamond shapes. An imitation topaz was stuck on at the centre of each oval. There is one other known dial by the same dial maker using an imitation topaz, but no other dial maker seems to have used this decoration

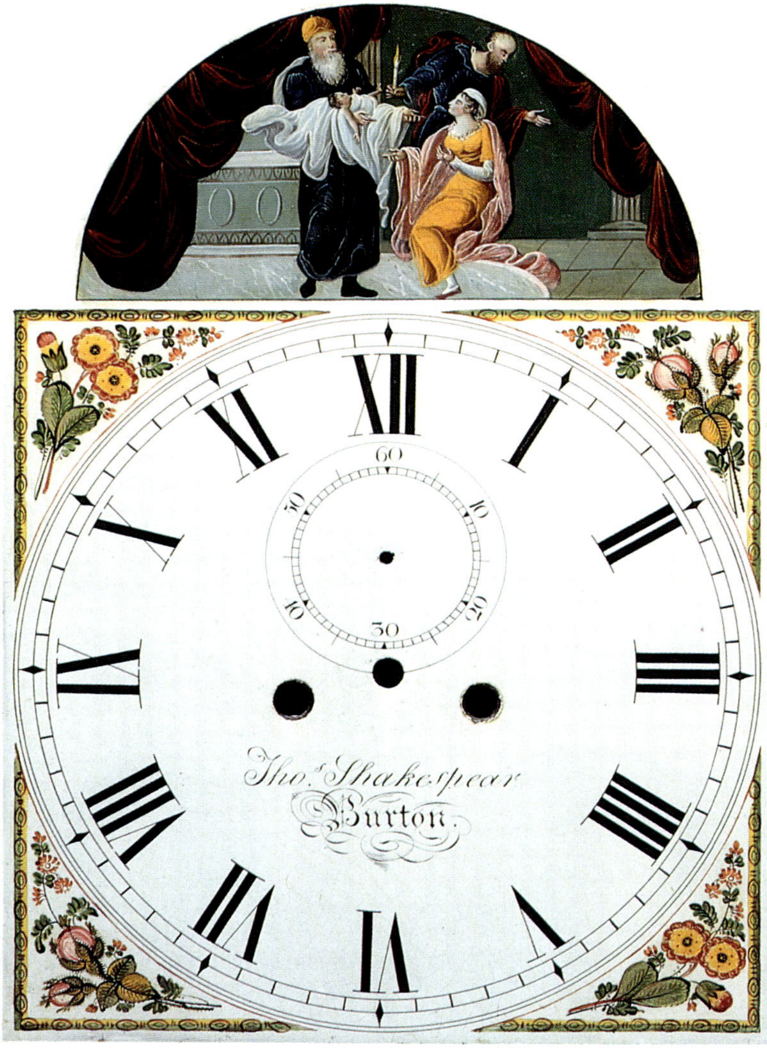

(xii) Thomas Shakespear of Burton upon Trent 1828–35. A Walker and Hughes of Birmingham dial *c.* 1825–30 with a break arch scene taken from 'The Presentation at the Temple' by Reubens. The numerals have reverted to the roman type but without any minute numbers. Very elaborate scrolling around 'Burton' is typical of Walker and Hughes dials

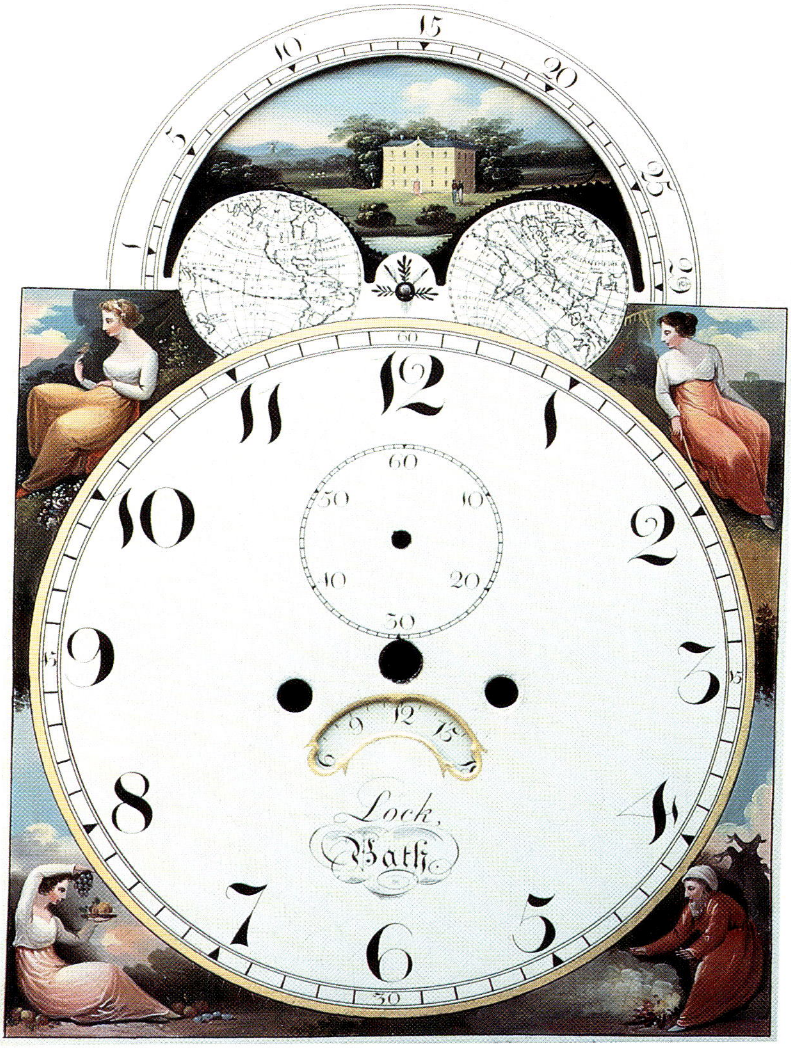

(xiii) James Lock of Bath, 1780–1830. Another Walker and Hughes of Birmingham dial, 1815–20, with corners representing the four seasons. The ladies are well painted, but 'Winter' lower right has a feminine profile with a beard added so as to represent a man. There are two other dials by Walker and Hughes with bearded ladies

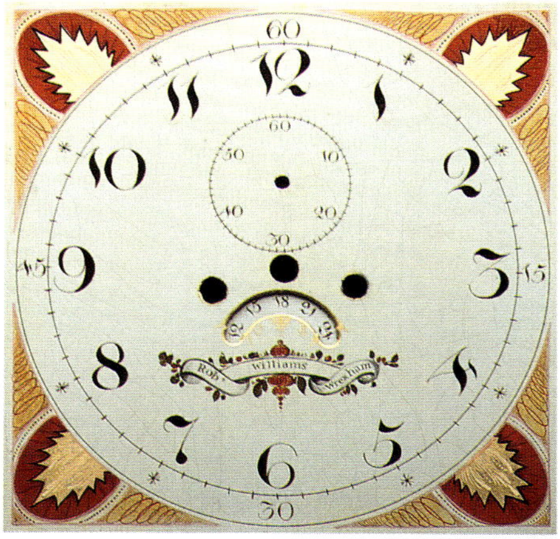

(xiv) Robert Williams of Wrexham, Clwyd, listed as working 1822–44 but this is obviously a pre-1822 dial. It was made by Finnemore of Birmingham *c.* 1812 and features the star and dart corners first used by Walker and Finnemore. The decorated scroll containing the name is reminiscent of Wilson dials *c.* 1800

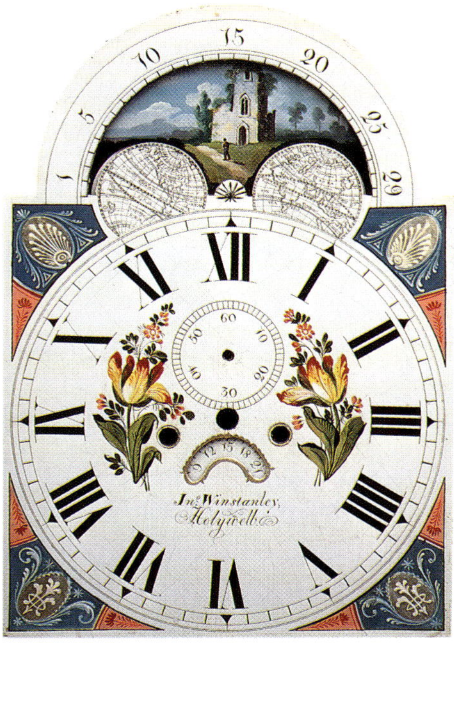

(xv) John Winstanley of Holywell, 1791–1840. This is a Hobson of Birmingham dial *c.* 1830–5 where the corners are very bright indeed and seem unrelated to the fancy tulips in the centre. The roman numerals are heavy and the minute band very wide with extended half diamonds

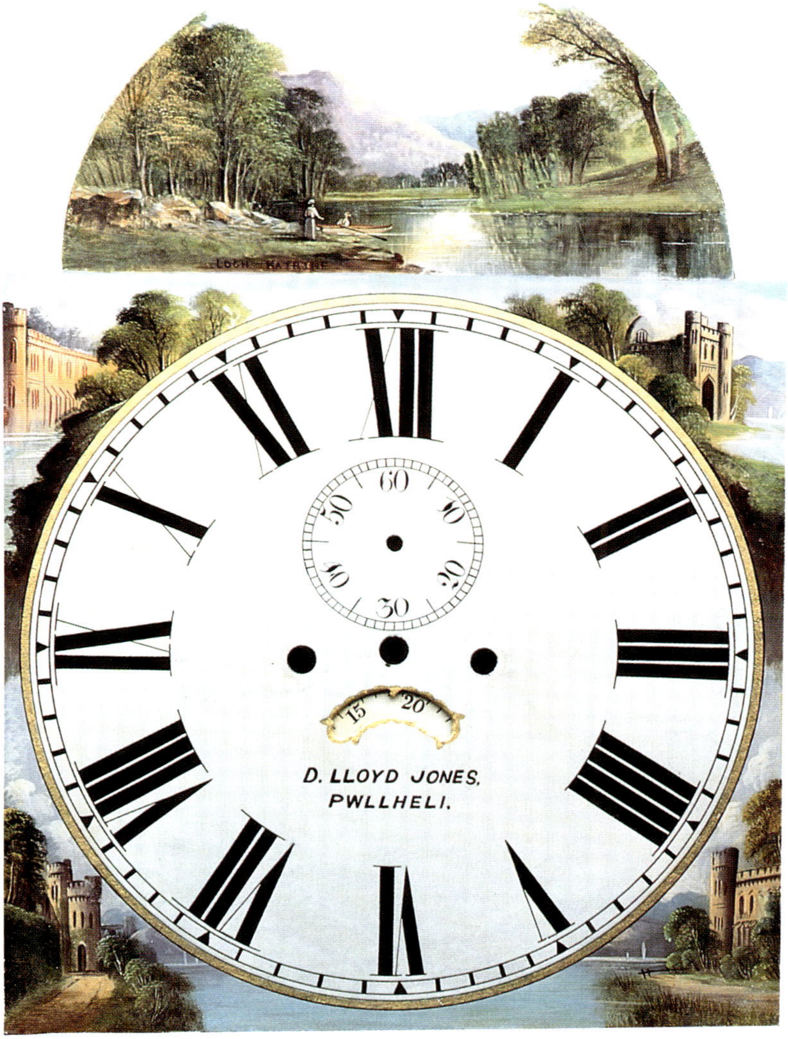

(xvi) D. Lloyd Jones of Pwllheli, 1860–70. An unknown Birmingham dial maker produced this well painted dial in a period when many dials were carelessly done. The scene in the break arch is Loch Katrine. The clock maker's name and town are done in the sans serif style which began to be used during the period 1845–50

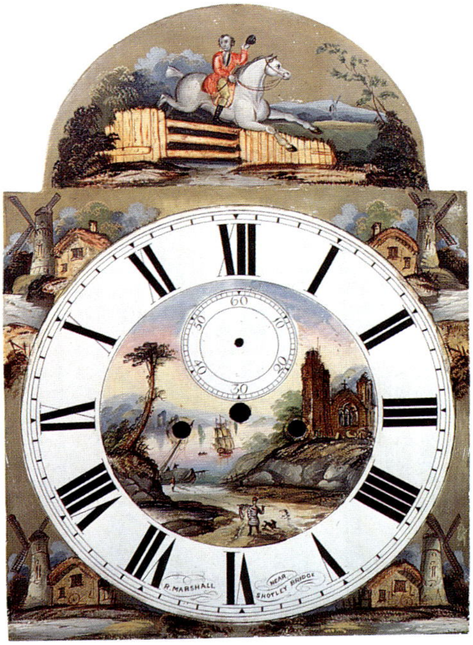

(xvii) Robert Marshall near Shotley Bridge, near Gateshead, 1848–58. Dial maker here is unknown though probably Birmingham *c.* 1850. There is a bronze powder background to the corners and break arch with gold or silvery leaf for highlights. The centre of the dial contains a fantastic scene with a sunset coloured sky which was popular around the 1840–50 period

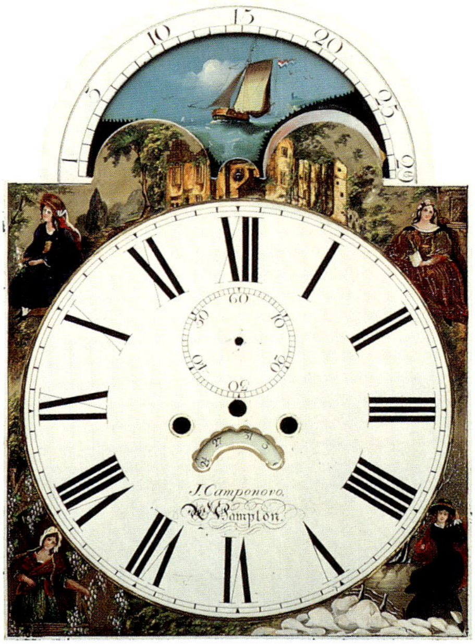

(xviii) J. Camponovo of Wolverhampton is an unlisted clock maker. The dial maker is also unknown but this is definitely Birmingham work 1840–50. The four seasons ladies have bronze powder backgrounds and the scenery at the top of the dial continues all over the hemispheres. The dress of the lady in the top right is vaguely medieval

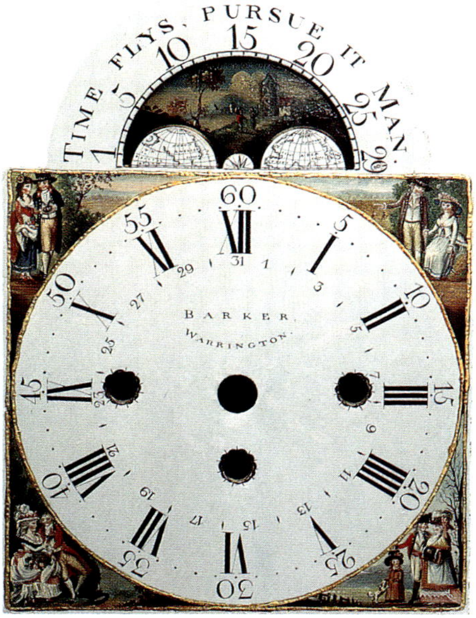

(xix) Barker of Warrington and Wigan probably 1780. A Manchester-made three-train bracket clock dial based on brass not iron with centre date hand and decorated winding holes. The corners represent the four seasons and the four stages of love, probably inspired by popular prints of the period. The rudimentary maps were hand drawn which is usual with all Manchester-made dials

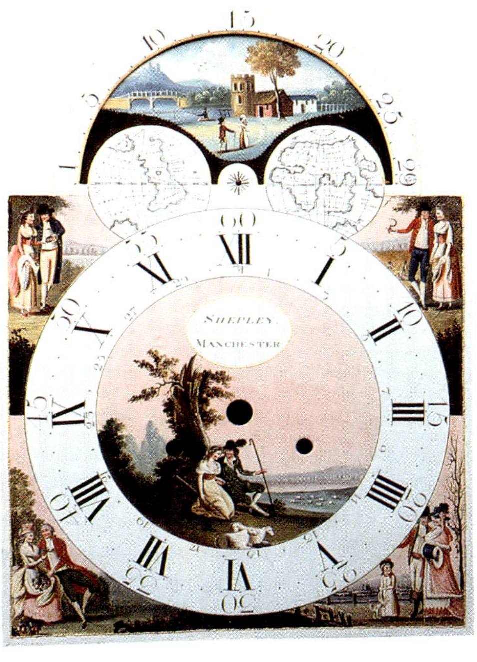

(xx) Edward Shepley of Manchester 1788–1800. Another Manchester-made dial but based on conventional wrought iron and made for a longcase clock. The corners are very similar to the Barker dial and the same artist must have painted both dials. The curious hand drawn maps can be seen in greater detail on this dial

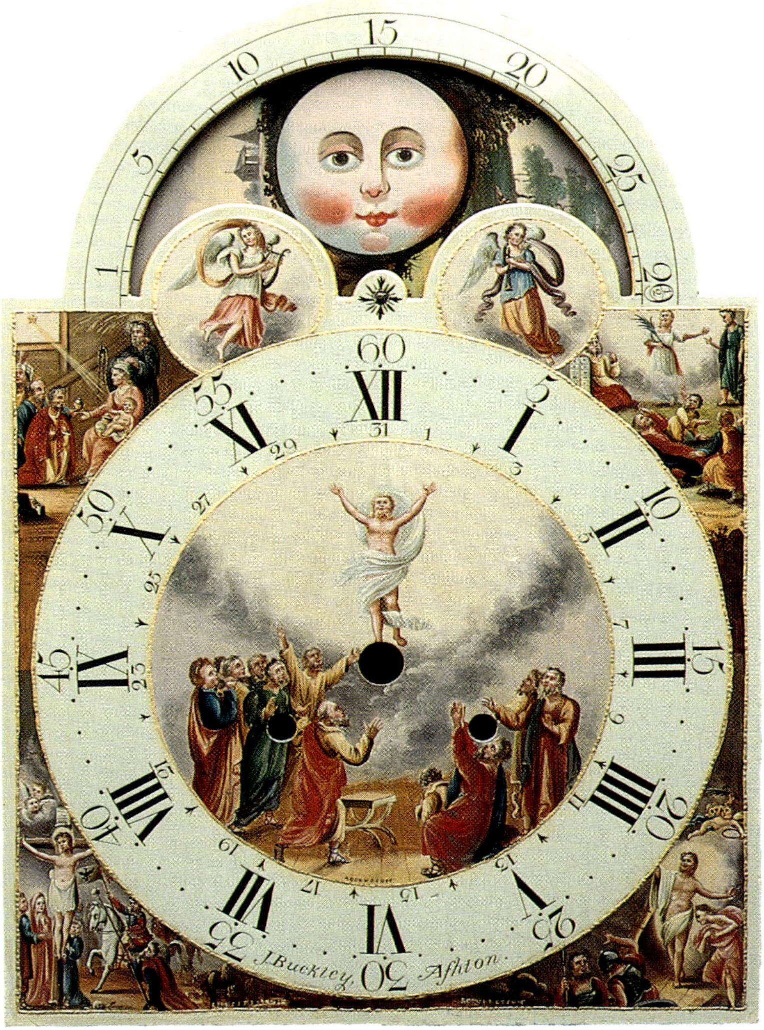

(xxi) John Buckley of Ashton-under-Lyne. The only date given is *c*. 1820. Yet another and probably later (1800–15) Manchester-made dial where religious themes are used in every possible place. Two other religious Manchester-made dials of this type are known. It is possible that the Buckley dial was a special order

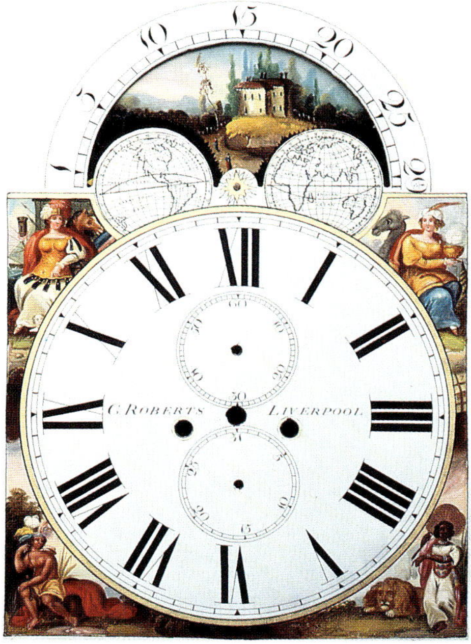

(xxii) George Roberts of Liverpool and Birkenhead. Poston and Woldnough of Liverpool made this dial *c*. 1835–45. The four continents are represented in the corners

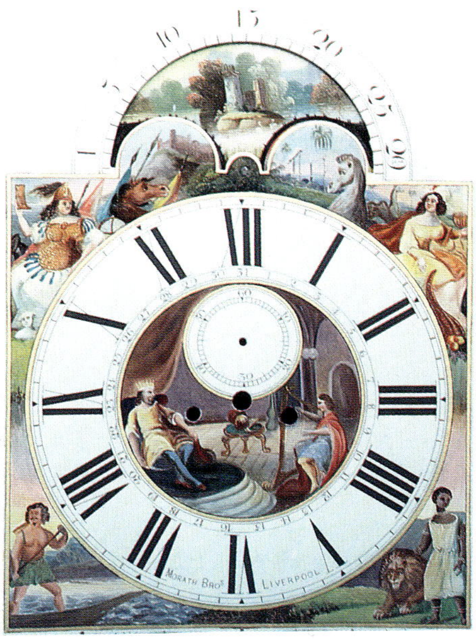

(xxiii) Fedele Morath and Brothers, Liverpool, 1848–51. The dial maker is unknown but it is possibly Liverpool made. One of the corners of the four continents has a most unusual Native American wearing a short goatee beard. The centre painting is of David playing his harp to King Saul

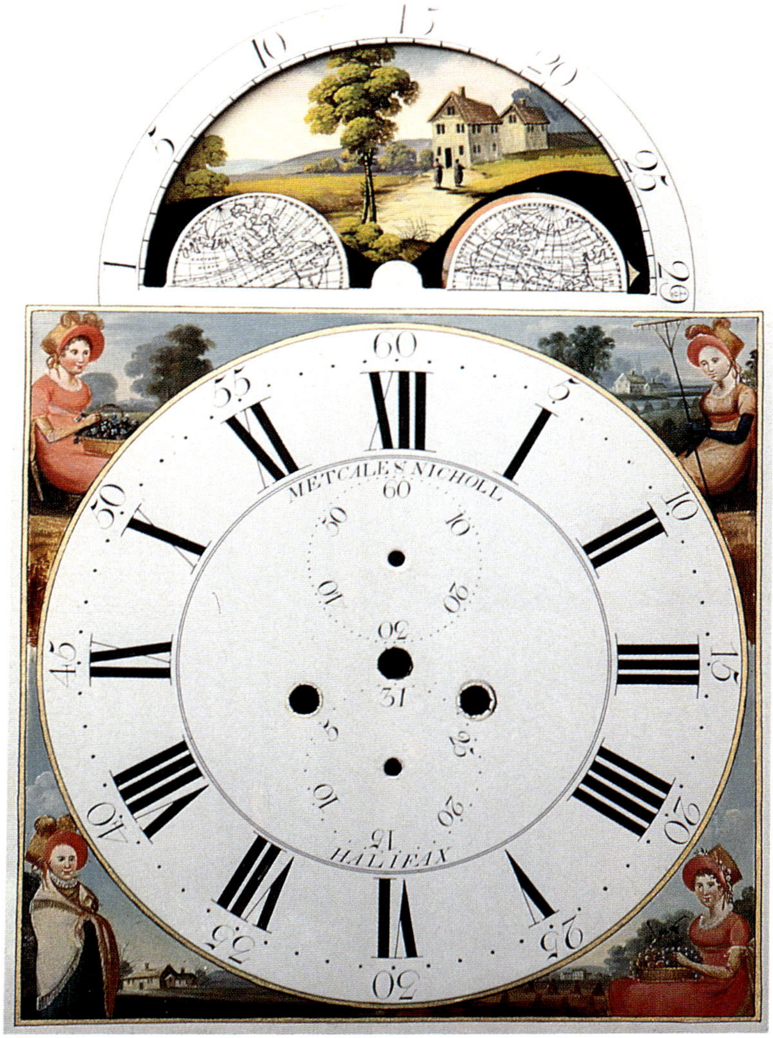

(xxiv) Metcale (Metcalf(e)) and Nicholl of Halifax, *c.* 1800. This dial was made by Whitaker or Whitaker and Shreve of Halifax. Though it was unmarked with any dial maker's name the style is definitely Halifax. The girls depicting the four seasons include one in a pink dress often seen on Halifax dials. The girl on the top right has elegant Prussian blue gloves for haymaking

3 Other Birmingham Dial Makers

ANALYSIS OF STYLE CHANGES 1800–1870

Before starting a general survey of Birmingham and other dial makers whose work has come into the author's workshop, it might be useful to note some of the changes in both graphics and subject matter. As seen in both Wilson's and Osborne's work, around 1800 the graphics changed from Roman numerals for the hours with full Arabic numerals denoting the minutes to Arabic numerals for everything. The author's theory is that the earliest of these 'Double Arabs' had straight verticals in the ones, fours etc. They may or may not have been 'tumbling' i.e. going radially, reversing the aspect between numbers three and nine. The later Arabic numerals became flared, which could have been considered more graceful. The author has also seen both styles of numerals on the same dial: straight numerals for the hours and minutes with the flared type for date and seconds. As the century progressed the minute numerals were put on the quarters only, with some sort of marking in between.

Various dial makers seem to have had their favourite marks, ranging from fleur de lys, diamonds or a sort of spider, but the Walker and Hughes trident was the most distinctive and was not used by any other maker, though that firm sometimes did use other marks. The size of these Arabic quarter numerals gradually shrank, and towards 1820 the dial makers reverted to Roman numerals with Arabic quarter numerals. All of this was subject to the wishes of the customer, so there were dials of 1815 which still used the earlier type of graphics. Around 1820 the minute numerals tended to disappear completely. The dotted minute markings on the earlier full Roman and Arabic dials were often replaced by either a plain circle with short bars through it or two concentric circles with either bars or dots or both included in the circles, as well as diamonds or half diamonds to denote the minutes. The minute band formed by the two concentric circles could be quite wide, especially in the 1830s and 1840s when the Roman numerals became very heavy indeed. There were, however, some Roman numerals of this same period and even later which were quite tall and slender. A very few clock dials of 1840–50 retained the full Arabic minutes, but the style of these minutes was far heavier than the elegant pre-1800 ones. Often in the cheaper 30-hour dials the main Arabic numerals, when used, were very heavy and bold.

One graphic oddity occurred where the Roman numerals were set vertically instead of the usual radial type. This was used by Eli Felton of Birmingham (1808–31) in a square dial which has Toleman, Carnarvon (1795–1840) on it, Arabic numerals for the minutes 15, 30, 45, 60, as well as dots and oak leaf interim motifs. The corners have a green background with flowers. Another dial of this type, which may also have been made by Felton, is a break arch illustrated in Brian Loomes's *White Dial Clocks*, p. 106. It is probable that this style of graphics was short-lived in the period 1808–10.

Another graphic rarity was a very late dial (*c.* 1860) made for John Griffith of Carnarvon with a fully-painted centre and large Roman numerals in black but with the 12, 3, 6 and 9 done in bright red. This was quite original, though the dial maker is unknown.

Lettering style for the name and town varied a great deal, but in general the earliest dials had simple lettering. There were, of course, exceptions to this, especially when the lettering was put on locally after the dial was made. After 1800 the more elaborate styles came into use, and often the later the dial the more complicated the scrolling around the name and town. The most extreme form of this elaboration was used in the 1830s and 1840s, when it was so overdone that it is often hard to decipher. There was a reaction to all this, however, and in 1845–50 a much simpler lettering, without serifs, was used.

The decorative parts of the dial altered as well as the graphics. The changes from the gesso-framed fruit or flowers in the corners to the use of the stylized fan or shell motif have already been noted. It is hard to be absolutely definite about which decorations were popular at any given time, because dial makers did vary their products. Samuel Deacon of Barton certainly specified just what he wanted in graphics, as well as decoration, in a letter of 1805: 'written figgers' were asked for, probably meaning Arabic not Roman, and some 13″ arched 30-hour dials were to have 'figgers in arch, one with an urn'.

With regard to corner painting, flowers and fruit done in various ways were used from the first to the last of the painted dials, a period of roughly a hundred years. The same holds true for birds, but it is in the manner of this painting that the dial can be roughly dated. The Osborne and Wilson birds were painted in some detail, but the late birds, such as those done by Finnemore, are much more sketchy. Shells and fan corners seem not to have been used much after 1820. Ruins, castles and cottages which first appeared in the centres of some late Wilson dials (*c.* 1809) were used very frequently from 1815 until the end of the painted dial in 1870. Geometrics were used at any time after 1800, especially because they were easy to do and therefore cheap. The four seasons in various modes of dress continued well into the 1850s, and the dress style, if nothing else, can help to date these dials. The four virtues, apostles and other themes were not quite so popular.

The four continents, which first appeared around 1815–20, continued into the late 1860s. Many of these paintings of the four continents appear remarkably similar, and it could be that they were based on prints of the period. The misconception that American Indians wore ostrich feathers on their heads seems to have persisted for a long time. Anchors were first used around 1800 and continued for about fifteen years, but the urns that were so popular from 1790–1815 did not continue much after 1820. The use of animals and birds other than the Wilson/Osborne type of bird started about 1820, and they were done in a naturalistic style, often with a background unlike the fantastic birds seen earlier. Animals and birds were uncommon after 1860.

As far as is known, only one dial maker – Walker and Hughes – used fake 'jewels' in both corner and break arch decoration. These were small topaz-coloured glass, glued in by a black pitch-like substance in the centre of a painted oval. Another unknown dial maker produced a dial for Thomas Ross of Hull (1790–1812) with what appeared to be pressed brass naval motifs with a foil backing in the centre. These were definitely original to the dial. Anything of this sort could be considered rare, though Brian Loomes has included in his *White Dial Clocks* (p. 112) a photograph of one dial with convex beads stuck on in the centre of an urn. Unfortunately, the dial maker is not named, though it could have been Walker and Hughes.

The break arch area of the dial might have been painted with almost anything. Flowers were often used up to 1820, as were sea shells. Ships appeared around 1800, and these could have been used in the dial centre as well, but after 1825 ships were unusual. Single figures not contained in an oval scene began to be used in the break arch about 1805, but these gradually gave way to the fully painted scene around 1815. From 1815–50 the break arch often contained a hunting scene, and a favourite one showed the horse with all four legs off the ground. Romantic views with ruins and castles began to be popular as a break arch subject around 1815, and this continued up to 1860. Farm scenes, often with animals, were used at about the same period. Some break arch designs were copied from popular prints or paintings, so that the same theme could turn up in different versions. W. Francis of Birmingham produced two dials for Wilkinson of Leicester, both of which featured 'Pig Sticking in India', but the painting had definitely been done by different artists. Named views such as 'Durham

Cathedral' or 'Lake Garda' were uncommon, but they *were* done, though the title was sometimes very small and obscured in the painting. Topical events such as the death of Nelson, a particular Royal visit or even a prizefight could also feature in the break arch and sometimes this can date a dial to a particular year, though the 'Death of Nelson' continued as a subject for a considerable time after the event.

Gold leaf had been used in various ways from the beginning of the painted dial, but silver or possibly tin leaf started to be used around the 1840s. It was cheaper, and gave a beautifully soft effect, especially in religious break arch scenes, such as 'Moses Receiving the Ten Commandments' or 'The Last Supper'. Silver or tin leaf could be used in conjunction with the more expensive gold leaf, but the special effect of the silvery leaf tended to be lost when it was overlaid with yellow translucent varnish in order to imitate the gold leaf. At the same time, in the 1840s, a very colourful treatment of painting the sky was popular. It was done in sunset colours and needed a skilled hand in order to make it effective. Another variation, slightly later, was the 'gold' or bronze powder background to the break arch figure or animal painting, and this could also have formed the background to the corner paintings. Invariably this bronze powder has darkened with time, but when new the effect must have been very brilliant, especially when combined with silvery leaf.

Another stylistic innovation in the 1850s' break arch was the painting of a false moon, complete with numerals around the arch, which was to give the impression that the clock had moon works. Some of these false moons were very crudely painted and would not have fooled anyone. It could be considered that a break arch was a permanent part of the dial, but that is not always the case. One break arch Finnemore dial was transformed into a square one soon after manufacture (*c.* 1830) by the simple procedure of cutting off the top. This left a rather stepped edge on the top, and was presumably done by the clock maker Joseph Hollyoak of Smockington. Another square dial had break arch added, using three riveted strips to hold it in place on to the main dial. There had been no attempt to hide the rivet holes or the longitudinal crack across the top. The style of the break arch looked slightly later than the main dial, which had been made for T. Leeming of Bentham between 1825–35.

Moon dials could vary enormously in the way that they were painted, but all had to have the necessary moon face. Certain artists can almost be identified by their moon faces, as mentioned before with some of the Osborne moons. One method of grouping together some late dials where the dial maker was unknown is to study the moon paintings. The early period (1772–1810) moon dials were always well painted, and the landscapes done with a considerable degree of expertise. However, as time went on the painting of the moon dial became, like the main dial, more and more slapdash, and some of the 1860s moon faces were almost grotesque.

The maps used on the hemispheres of a break arch moon type of dial are really a separate subject in themselves. Both Wilson and Osborne had various versions of the maps, depending on the size of the dial and the period. The maps of both these makers seem to have been periodically updated. Until about 1820 other dial makers usually had their own particular types of maps but some dial makers, such as Francis Byrne, Ashwin and Co., and the Nicholas brothers, used the same transfer prints. In about 1820 the outside marking of latitude degrees was first used, and this continued to be used by some makers until 1870. Map transfers were tricky to put on, so in 1825–30 another variation in the treatment of the hemispheres appeared. A small globe was sketchily painted and the dove of peace, carrying an olive branch, flew above this, often with a sunray effect background. A gold leaf border, either plain or scalloped, on the edge of the hemispheres added to the decoration, but this was not always used. Another style had single figures, with or without scenery in these hemispheres, and some used the scenery alone. These fully painted hemispheres were used right up to 1870.

Centre dial painting appeared around 1790. This included a scene, or group of figures on the centre of the dial, as distinct from just two birds or bunches of flowers, shells or ships. It was not a common thing to be found on a dial as it was obviously more expensive, but centre painting was used until 1860. In some cases there was a date aperture, but other dials could have had a date and/or seconds circle which tended to interrupt the design. Occasionally, two oval miniature-style portraits were included in a centre dial painting of a land or seascape. These portraits were usually named. All sorts of subjects were used, rather like the break arch paintings. The early pre-1800 centre paintings tended to be pastoral scenes and similar to the break arch work. Seascapes with ships came into fashion about 1800. Some military scenes began to appear at this time as well, and the death of Nelson produced a veritable flood of commemorative dials. Around 1825 fully painted rural pictures were used, as well as hunting scenes. Religious themes appeared in the dial centres *c.* 1835–40, and these were sporadically used until 1860.

Automata have had a long history of being incorporated into clock dials, and they turn up in nearly every period, except in the earliest 1772–80 dials. Automata were the most expensive of all dials, especially in the 'Adam and Eve to Move, with Moving Serpent'. All sorts of themes were used. Boy and girl on a see-saw, a game of Battledore and Shuttlecock, Harlequin and Columbine, man sawing wood, musicians playing, Father Time, rocking ship and a moving swan all could feature in automata, but the most popular by far seems to have been the rocking

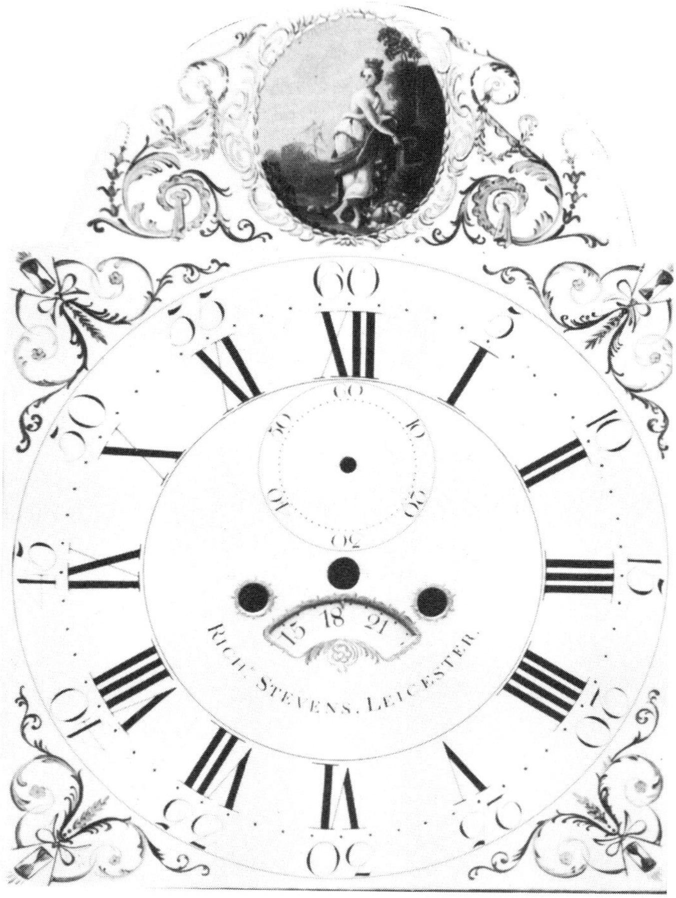

38 Early 1780s dial made by Francis Byrne of Birmingham for Richard Stevens of Leicester. The corners feature scythe and hourglass, while the winding holes are gilded and the date aperture is of the square-ended type. See page 71

ship, closely followed by Father Time. The method of painting these automata, even in the pre-1800 era, could be considered a bit crude but the thin metal of which the moving figures were made meant the paint flaked off rather easily as the metal became bent. Many highly ingenious automata were devised, some of which, being insufficiently powered, tended to stop the clock. Thus in many cases the clock could be minus all of the mechanism to work the automata, as repairers in the old days did not seem to want to bother with these complications. A late and rather comic form of automata was the lion with rolling eyes used in the 1860s. The early apertures for automata were regular in shape, or else the scene was painted on to the main part of the dial and the moving figures stood proud of this. Later on, probably 1830–40, the apertures could be irregular with cut-outs resembling rocks and grottos usually associated with a rocking ship, or else a military scene with trees and a gun barrel as well as a lone cannon ball. One very unusual Walker and Hughes dial (*c.* 1815) was ahead of its time in using an irregular aperture for a rocking ship, but this was very rare in that period.

SURVEY OF BIRMINGHAM DIAL MAKERS' PRODUCTS

There now follows an overall survey of those Birmingham dial makers whose work is known to the author, noted chronologically wherever possible. Birmingham continued to be the main centre for this trade during the whole period, and a great number of firms appeared on the scene after the deaths of Wilson and Osborne. Some of these makers seem to have produced very little, or else they folded up after only a few years. Brian Loomes in his book *White Dial Clocks* has produced a very good list of all recorded dial makers, dial painters and anyone else associated with this trade but, obviously, since that time other makers have come to light.

Although the author has not dealt with all of the makers on this list, in seventeen years a pretty representative selection of dial makers' products have come to the workshop. Nothing on a subject as broad as this can be stated to be truly definitive, because the more that is known about dial makers, the more examples turn up.

FRANCIS BYRNE

Francis Byrne (1780–1810) was obviously working in Birmingham at the same time as Wilson and Osborne, and some of his dials could at first glance be thought to be Wilson ones. However, on close examination they differ considerably. The earliest and most unique style is shown in this dial (**Illustration No. 38**) made for Richard Stevens of Leicester. The date is given as 1791, but it is certain that this particular dial was made in the 1780s. The date aperture is of the square-ended type, and the winding holes are gilded. The goddess in the break arch is Cybele, mother goddess of Phrygia. The scythe and swag corners, done in delicate shades of pink and blue, were not used by any other maker. A somewhat similar dial with scythe and swag corners is found on p. 217 of Brian Loomes's *Grandfather Clocks and their Cases* but the dial was attributed, though unchecked, to Osborne, and the corners are supposed to be gilded. Although there appear to be no grommets, it is possible that the restorer did not see any evidence of gilding or, if unrestored, the gilding has worn off – a quite common occurrence.

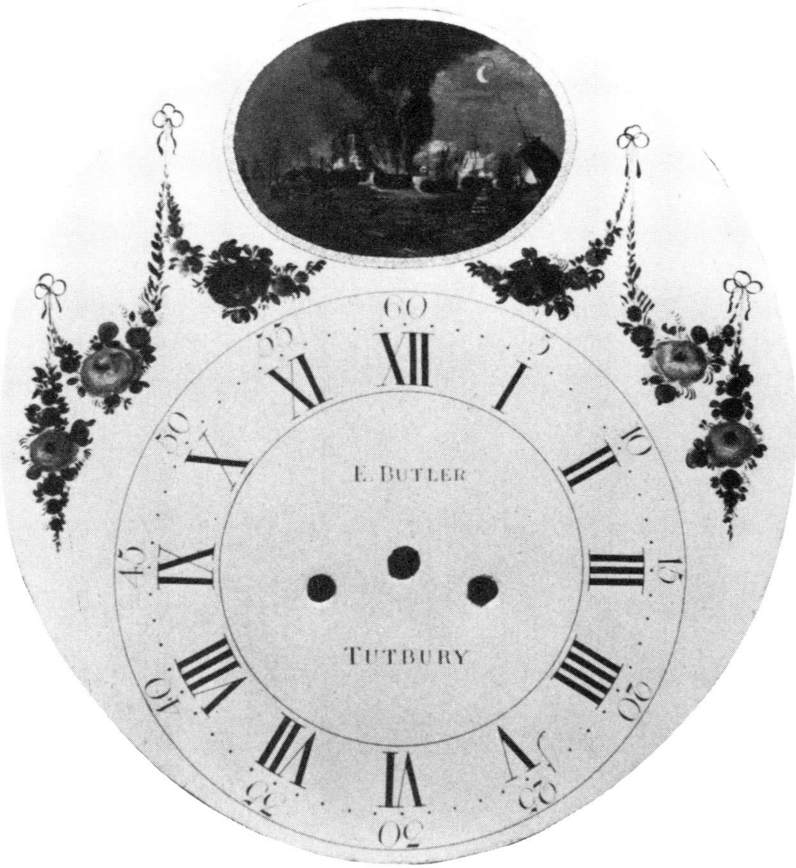

39 This is a very large oval Byrne dial showing the Battle of the Nile, August, 1798 as well as the bold painted flowers typical of this dial maker. It was made for Edward Butler of Tutbury. See page 74

Byrne used fine gilded gesso work well into the 1800s. He sometimes used nothing but gesso to decorate a dial, with small flower sprigs in roughly the positions of the four rivetted feet, but most often the gesso was used for scrolling in the break arch that surrounded a well-painted vignette or else gesso was used as framing for floral corners. Gesso also turns up in fan corners or those using a circle set into the corner. Byrne's painting of fruit and flowers can only be characterized as exuberant and less controlled than Wilson or Osborne. Even the birds, when used, are distinctive and often have a weeping willow sort of branch above them.

Another interesting point is that nearly all of the ladies depicted on Byrne dials seem to have very heavy thighs and relatively small heads. The transfer maps used on dials with a moon are also very different to those used by Wilson and Osborne. The lines of latitude and longitude were far closer together, and a great deal of detail was included, such as 'Cooks' Track'. As Byrne and Nicholas used the same transfers, the illustration of a Nicholas dial (**Illustration No. 40**) shows the type of map involved. However, the Byrne maps were not complete hemispheres because they were cut off at the chapter ring. One of the

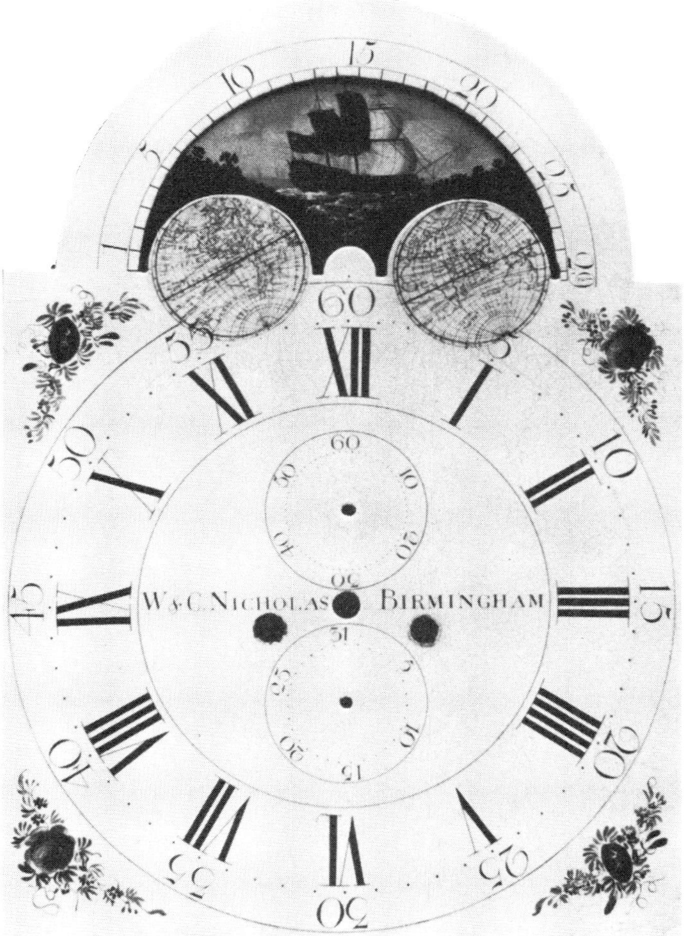

40 The Nicholas brothers of Birmingham made this rare dial with complete and detailed world maps. The dial is *c.* 1794 and the graphics are overlaid by the maps. See pages 73 and 75

most important ways of recognizing a Byrne dial, especially a 30-hour one or one with long feet and no false plate, is by the number painted in what was originally white on the back of the dial. These numbers are usually quite large, though not all Byrne dials have them. They appear to run consecutively: an all-gesso decorated dial of about 1785–90 had 173 on it, and another of 1805–10 had 1281 on the back. The author believes this numbering gives a very important clue to the quantity of regular production dials made by Byrne. Special orders do not appear to have been numbered, such as this large oval dial (**Illustration No. 39**) made for Edward Butler of Tutbury (1795–1828). The Battle of the Nile, 1 August 1798, can be seen clearly at the top of the dial, and probably this dial was produced to commemorate the event. It is a rather wide oval, and the floral swags are typical of Byrne. He must have been quite highly regarded as a dial maker, because the famous clock maker and inventor Edward Massey chose him in 1804 to make the dial for a magnificent large clock which can be seen in the Great Hall of the National Trust Property, Dudmaston. The dial (**Colour Illustration (vii)**) has a large-thighed Britannia in the break arch, with moss roses either side, together with fan corners in green. The numerals are Arabic, but only mark the quarters. As far as is known Byrne only used one type of cast iron false plate, with 'Birmingham' in capitals at the top, and 'F. Byrne' also in capitals at the bottom.

ASHWIN & CO.

Ashwin and Co. has only one date (1790) and no known address, but he definitely was a Birmingham maker, though few of his dials seem to exist. He also could do elaborate gesso work, and his flower corners were very similar to those of Wilson. As mentioned before, he used the same transfer maps as Byrne. The author believes he was only in business for a short time, as the three Ashwin dials to come in for restoration all have the Roman and full Arabic numerals usually seen in the pre-1800 period. Also, the cast iron false plates with only 'Ashwin & Co' in large capitals are very sharp indeed, usually indicative of a short production run. Samuel Deacon of Barton used one of his 8-day dials for a 30-hour clock with a dummy winding square attached to the false plate on one side, and what had been the seconds hole at the top of the dial was neatly plugged with lead and then painted over with an elegant urn. The corners of this particular dial were flowers framed in gesso. It is interesting to note that although Ashwin dials seem rare in this country, they certainly were exported to the United States. Two examples, both with gesso work, are in the Mabel Brady Garvan collection at Yale University. Both are well illustrated on p. 45 and p. 165 of *The American Clock, 1725–1865* by Edwin A. Battison and Patricia E. Kane.

HIPKISS & HARROLD

Hipkiss & Harrold (1797–1805) was yet another firm that did not seem to produce many dials, though all of the dials seen by the author from this maker are of high quality. The only really distinctive thing about Hipkiss and Harrold dials was the use of very fancy parrot tulips in the corners surrounded by gilded gesso framing. John Deacon of Leicester, son of Samuel Deacon, used a Hipkiss and Harrold dial of this type for a clock with moon dial and centre date numerals *c.* 1798. The cast iron false plate was of the flowing script type using the long 'S'.

WILLIAM, CALEB & JOSHUA NICHOLAS

The Nicholas brothers were an interesting Birmingham family because, although they made clocks and watches, they also produced dials both for their own use and for sale to others. Caleb (1787–1843) worked with William the dial maker from 1785–7. Caleb had no less than six different addresses. William (1785–1825) only seemed to have moved three times, from Park Street (1785–6) to 6 Smallbrook Street (1787–1824) and lastly to 9 Smallbrook Street in 1825. Joshua (1797–1801) worked at 14 Bull Street. Presumably William carried on his dial making business all of his working life, but his dials can be difficult to recognize if a named false plate was not used. A very elaborate (**Illustration No. 3**) false plate certainly was used for the dial in **Illustration No. 40**. It is interesting to note that, although the false plate had 'W.C. & J.' on it, the dial belongs to the 1785–7 period. It is most unusual in the use of the whole globes, and this type of transfer map, as mentioned before, was used by Ashwin and Byrne, though in a more conventional way. Another very interesting dial has turned up, marked 'W.C. & J. Nicholas, Birmingham' on the dial face, but using a very plain cast iron false plate with only the number '243' punched into it. It is quite a large dial, and thus unlikely to be very early. This dial has full Roman and Arabic numerals, and in the break arch is an oval vignette, well painted, of a seated lady with a flute in a pastoral scene and a small dog, which has a remarkably human face, at her feet. The floral corners impinge quite a lot on the chapter ring. A slightly later Nicholas dial was made for H. Knight of Stone which had floral gesso framed corners, two Wilson-type birds in the centre and all Arabic tumbling numerals and the elaborate style of false plate. The latest recognizably Nicholas dial was one with the four seasons in the corners, straight set Arabic numerals and quarters with a scalloped gold outside edge to the chapter ring (**Colour Illustration (viii)**). It was marked 'W. Nicholas, Birmingham' on the dial face yet the plain false plate had 'C. & J.

Nicholas' on it. This particular dial had a moon dial, and the maps are of a most peculiar variety with very strange North and South Americas, as well as outside latitude markings. This seems to be the first use, presumably pre-1820, of maps with these markings, and the identical maps were used by W. Francis (1810–31) for a four-apostles dial with no clock maker's name and straight set Arabic numerals for the hours. The same maps turned up on a much later dial with an unmarked sheet iron false plate made for H. M. & S. Huntington of Chester *c.* 1845. It is to be hoped that more Nicholas dials will eventually turn up as they could span quite a long period of time.

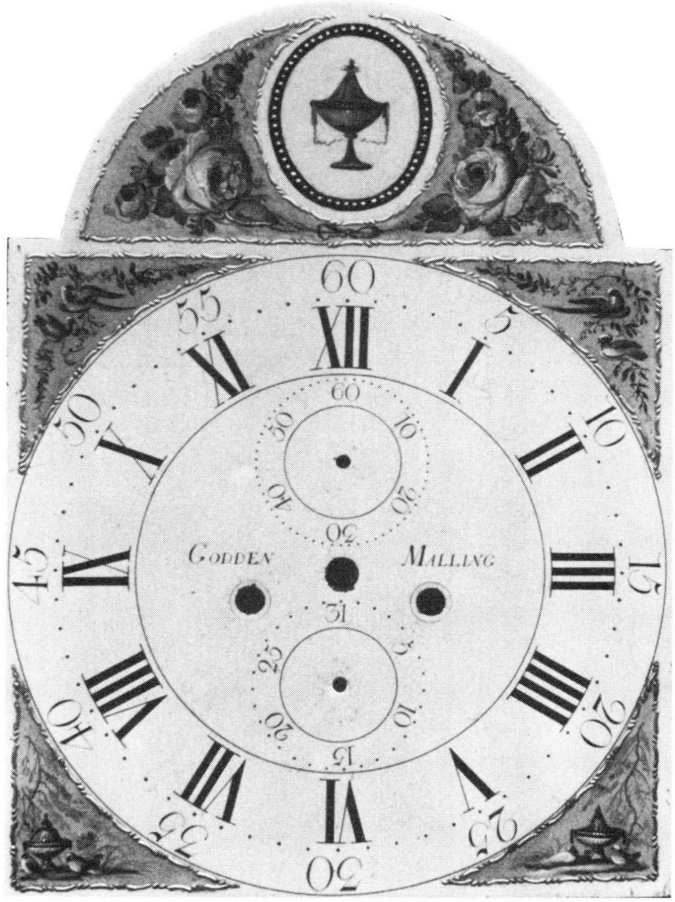

41 An Owen of Birmingham dial of great elegance made for John Godden of Malling *c.* 1800. See page 77

OWEN & PRICE

Owen & Price of Lower Priory (1800–02) produced dials similar to those of Wilson with simple floral corners and a bird in the centre. Not many of these Owen and Price dials seem to exist due to the short life of this firm.

EDWARD OWEN

Edward Owen (1803–21) followed on working alone in Lower Priory and was quite prolific. The earliest of the Owen dials seen by the author often had a very blue base paint. The corner motifs varied from gold leaf flowers to shells or fans. This elegant dial, (**Illustration No. 41**) made for John Godden of Malling about 1803, used a yellow ochre background to the pink flowers and birds. The decoration is reminiscent of the furnishing fabrics used at that period. Owen could also produce elaborate gesso work corners, or else Wilson-style floral corners with gesso framing. The birds when used are also similar to those of Wilson. Quite a few corner decorations were of the geometric style, often including quite a lot of gold leaf. The author believes he was the first dial maker to use the popular four continent corners. This dial (**Illustration No. 42**) was made for Edward Glaze of Bridgnorth (1770–1807). It is interesting to note that the so-called American Indian appears to be seated on a crouching Negro. It is probable that these four continent corners were roughly based on prints of the period, as mentioned before. The author has restored another Owen dial of this same type, obviously painted by the same artist, but there are slight variations in costume and background.

As far as is known, Owen dials were not originally exported to the United States, though doubtless some of them exist there due to the efforts of the antique trade. However, at least one spectacular special order dial was exported to Ireland (**Illustration No. 43**). James Colhoun and Son of Londonderry (1795–1833) must have had a customer about 1820 with a particular reason for this unusual decoration, especially in the break arch scene depicting the attack by Cossak (sic) on General Le Fevre. The corners represent the four parts of the British Isles, and the fully painted centre has Britannia in the middle with portraits of Collingwood and Nelson above. One of the last dials Owen made was one for William Green of Prescott (1819–34) with Roman numerals only, and rather silly looking sheep in the corners. The painting appears to have been done in rather a hurry, though this was an 8-day dial with a break arch moon.

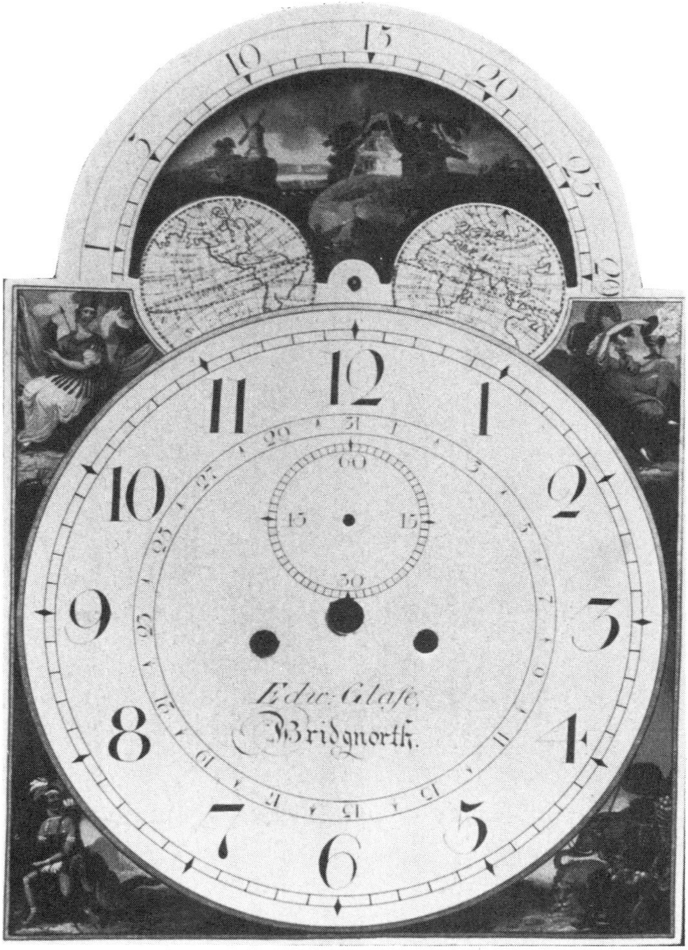

42 Owen dial made for Edward Glaze of Bridgnorth *c.* 1807 which shows early use of the four continents in the corners. Note also the centre date markings which were unusual for this part of the country. See page 77

NATHANIAL PORTER

Nathanial Porter (*c.* 1812) dials are very rare indeed. After the death of James Wilson on 3 April 1809 there must have been a great number of employees out of work. The only firm evidence of anyone trying to take over the business comes from one dial with a false plate (**Illustration No. 2**) which is marked 'N. Porter, late Wilson' on one side and 'Wilson' on the other side. The dial illustrated (**Illustration No. 44**) was made for William Knight of Stafford whose dates are unknown. The only date

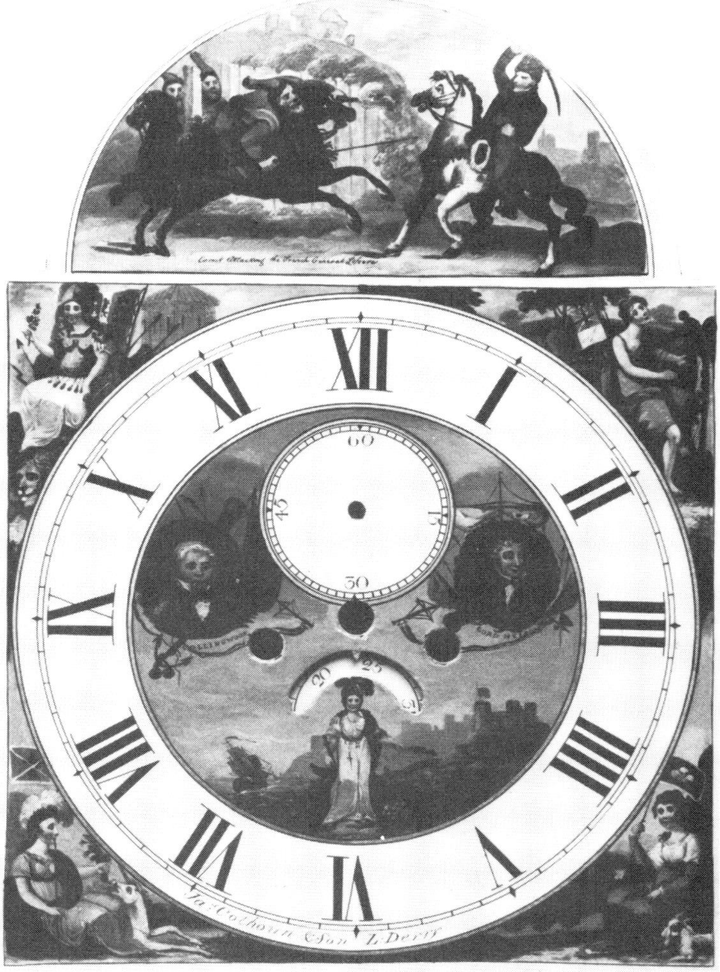

43 A very unusual Owen 'special order' dial made for James Colhoun of Londonderry *c.* 1820 featuring 'Attack by Cossak [sic] on General Le Fevre' in the break arch as well as Nelson and Collingwood superimposed on the fully painted centre. See page 77

for Porter, given by both Loomes and McKenna, is 1812, but of course this was taken from street directories of the period, so possibly Porter could have started his business a year or so earlier. Certainly the street address, Great Charles Street, was the same or near to the Wilson premises. This particular dial is very different from any Wilson product, though Wilson maps were used. Even the moon faces are quite different. This firm must have been a short lived one, with a small production.

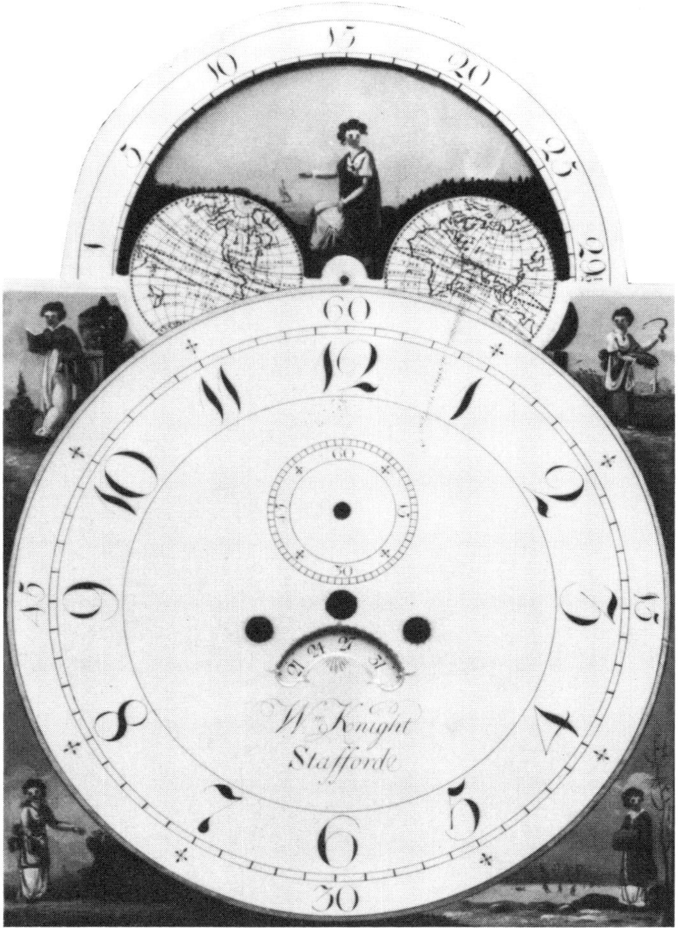

44 Very rare Nathaniel Porter dial. He was the short-lived successor to Wilson (see Illustration No. 2 for false plate). The corners show the four seasons. The moon dial is illustrated on page 206 (Illustration No. 128) and the maps are the same as those by Wilson. See pages 78 and 79

THOMAS KEELING

Thomas Keeling (1800–25), numbers 1 and later 15 Great Hampton Street, seems to have been in business a long time, but the author only knows of three dials definitely from him, though of course a lot of 30-hour or unmarked long footed dials could exist. One particular dial appears to have been quite early (*c.* 1808) and looks very like a Wilson one with birds in the centre, Roman and full Arabic numerals and gesso framed floral corners. It was made for William Roberts of Amlwch

(1808–25) and had the unusual feature of two small black scrolls just about where the bottom dial feet would have been rivetted in. These feet often caused cracking of the base paint, so this decoration could have been done in an effort to direct attention away from a possible fault. Another clock dial of about this period was used by Christian Hall of Litiz, Pennsylvania (1775–1848). It is illustrated on p. 104 of *Clockmakers of Lancaster County and Their Clocks, 1750–1850* by Stacy Wood Jr and Stephen E. Kramer III. This particular dial has flowers in the corners surrounded by gesso swags in a roughly round shape and the figure of a girl harvesting in the break arch. The third and definitely later dial was seen by the author in America at the Time Museum, Rockford, Illinois. It had been made for John Reed of Brampton, whose dates are unknown. The corners consisted of fine gesso in a circular shape with swags, and there were straight set Arabic numerals with quarters. In the break arch was a well painted seascape, with two rocking ships.

JONES & CARTER

Jones and Carter (*c.* 1810) is another firm whose work is rarely seen, though again unmarked dials from them may be in existence. The author has only restored one of their dials, made for John Winstanley of Holywell (1791–1840). It is an 8-day dial with maps and moon dial, tumbling Arabic numerals and quarters, together with small baskets of flowers in the corners. This dial was well made but not distinctive in any way. There must have been a number of small dial making firms that jumped on the bandwagon when the painted dial became so popular, but either they did not last very long or else they made few dials that could be definitely attributed to them because there were no marked false plates or date and moon dials. It is important to record these unusual dial makers, however obscure, in the hope that other examples of their work will turn up.

BURGESS

Burgess (1810–1815) was another of these so-called unknown dial makers that do not seem to be listed anywhere, but he certainly did exist as evidenced by one dial photographed by the author which was made for Francis Campbell of Oswestry (1822–41), probably earlier than 1822. It had flared tumbling Arabic numerals and quarters, geometric corners and a gold leafed balloon in the break arch. The cast iron false plate just announced 'Burgess' in large capital letters which were not very sharp. The dial appeared to be rather roughly painted. The one remarkable thing was the interior minute markings, which looked like small Christmas trees. Another Burgess dial came to the author's attention in the September 1989 issue of *Clocks* magazine, p. 57. It was made for Robert Fletcher of Chester (1800–34), and had large Roman numerals, Arabic quarter numerals with unusual v-shaped minute markings. There were two birds above centre, two small bunches of flowers below centre and fruit with flowers in the corners. This dial had a break arch moon, and what appeared to be very odd maps, though it is possible that this dial has been restored and the maps might not be original. The style of this dial was probably 1815–20. Certainly both dials do look like Birmingham work.

B. WRIGHT & CO.

B. Wright & Co. (1805–20) was also a firm whose dials are not often seen. The author has restored one very plain undecorated dial with full Roman numerals and Arabics made for Joshua Knight of Halstead, an unlisted clock maker, where the name was set into a circle in the break arch. Both the base paint and all the graphics were of exceptionally high quality, and this dial probably dates to about 1805. The earlier productions of this firm, up to 1815 or so, seem to be of the best quality. The standard of the art work was high and there are two dials on exhibition in the Newark Houses Museum in Leicester which illustrate this. One was made for John Chapman of Loughborough (1760–1840) and the other for Deacon of Leicester. Both have straight set Arabic numerals and quarters as well as nicely painted scenes in the break arch. Towards 1820 the standard appeared to deteriorate with regard to the artwork. Two dials exist from this later period. One was made for John Wilkinson of Leicester (1815–26) which had Roman numerals only, lightly painted convolvulus and primula corners with a farm scene in the break arch. The second dial was a break arch moon one, made for Francis Walker of Maryport, and its dates are given as 1828–48. The corners depict the four seasons, but the moon faces are too small for the hemispheres, and the dial had also Roman numerals only. The moon dial was not a replacement as it was stamped. There was another firm, Christopher Wright, whose dates are much later (1835–45) but dials from that firm had dial plates or sheet iron false plates stamped 'Wright, Birm', as opposed to Wright and Co, who used a cast iron false plate 'B. Wright & Co Birmingham'. The Christopher Wright dials will be commented on later in the text.

ELI FELTON

Eli Felton of 44 Weaman Street (1808–31) was also listed as a button maker. There still exist quite a number of his dials which show a great diversity of styles. The earliest ones were very similar to those of Wilson with floral corners and full Roman and Arabic numerals through geometric corners and flared tumbling Arabics as well as shells complete with seaweed in the corners and gold leaf roses in the break arch. One spectacular brilliantly coloured dial was made for John Parr of Liverpool (1814–34) with straight set Arabics and quarters, gold band on the outside of the chapter ring and the four seasons in the corners, all with bright blue backgrounds (**Colour Illustration (ix)**). The style of painting is very like the Porter dial, although the ladies have whiter faces. This dial has a moon dial where the moon's face has very red cheeks and a nose to match, which is typical of Felton. This clock is owned by Merseyside Museums and is at present at Prescot Museum.

KEMPSON & FELTON

Kempson & Felton (1812–15) worked in the same Weaman Street as Eli Felton, but it is not known how and if they were associated. The author has only restored three Kempson and Felton dials, which is not really enough to characterize a particular style. One, made for Grocott of Holywell, had flared tumbling Arabics and quarters, geometric corners using some gold leaf and a collection of shells, some with gold leaf, in the break arch. Another, made for John Wainwright of Nottingham (1779–1820), also had flared tumbling Arabics and quarters, with gold leaf roses in the corners, a wading bird above centre and a strange single fish egg contained in an octagon painted in the break arch. The third dial had Roman numerals with Arabic quarters, and was also made for John Parr of Liverpool. Gold roses again featured in the corners, and the moon dial was definitely painted by a different artist than was seen on the other Parr dial made by Felton.

WILLIAM FRANCIS

William Francis (1810–31) was quite a prolific dial maker. He worked at Thorp Street, Old Inkleys and 11 Smallbrook Street. His dials were used by the Deacon family of Leicestershire clock makers, and a note dated 1820 to the factor or supplier of dials in Birmingham stated that Deacon would not pay the same price for W. Francis dials as he did for Walker and Hughes ones. Francis certainly did export dials to Ireland from 1810–15. At this time he used quite a blue background paint, and this rather curious dial (**Illustration No. 45**) is an example. Originally this dial was supplied with a rocking ship in the break arch, but presumably this did not suit Campbell and Orr of Bangor, so these two figures were substituted. The cockhatted man threatens the other barefoot pleading figure with a short axe or hatchet, the arm going up and down with each swing of the pendulum. No one seems to have

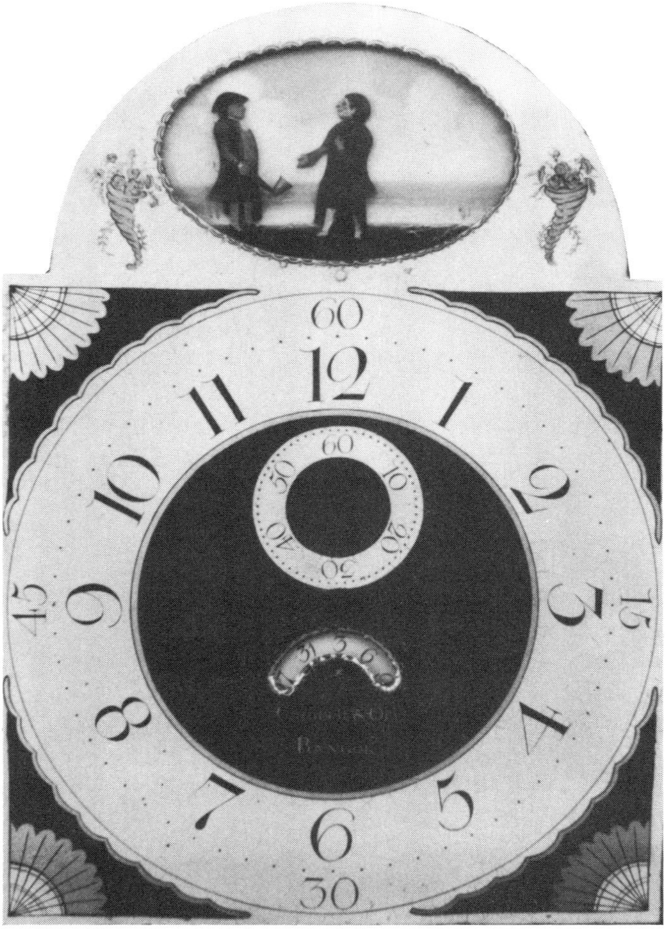

45 A curious W. Francis of Birmingham dial made for Campbell and Orr in Ireland *c.* 1815 which originally had a rocking ship in the break arch. This was changed, presumably by the clock maker, to the two male figures. See page 83

bothered about the fact that both men are standing in the middle of the sea.

During this period and possibly a bit later W. Francis cast iron false plates often had Wilson on the back of them. The name Wilson is usually very faint, indicating a worn pattern. This type of false plate generally had a scalloped inside edge on the top and bottom, but later false plates had scallops all around the outside edges. William Francis dials were done in a variety of styles. The earliest had the tumbling Arabic numerals with the blue ground paint and either geometric or gold leaf floral corners. Some of the more expensive dials, such as those featuring the four seasons or apostles, while retaining the Arabic numerals used a paler base paint. The artwork on these dials was well done, sometimes using very bright colours.

W. Francis moon dials could be iron, instead of the more usual brass, though this seemed to happen around 1820. When the graphics went over to all Roman numerals the popular castles and ruins tended to appear in the corners. One dial made for William Lines of Warwick (only date given 1828) used a bronze powder background for corner paintings of the four apostles, and the moon dial was iron. He did use the dished or convex dial from the late 1820s onwards, and these could be large. In fact one of the largest dials ever to come into the workshop was made by him: 15¼″ by 21¾″ made for Spencer of Bilston with flared tumbling Arabics, four continents and figures in the hemispheres. James Wilkinson of Leicester used W. Francis twice for dials decorated in the break arch with 'Pig Sticking in India', as mentioned before. Both these dials had Roman numerals and floral corners, but the artwork in the break arch was done by different artists. One of these dials was an 8-day one, and the other a 30-hour.

RICHARD BLOOD

At this point mention should be made of Richard Blood, who first turns up as a named dial artist used by one of the Deacons in 1813. Recently, the author restored a rather undistinguished looking dial with geometric corners and Roman numerals, made for James Hancock of Yeovil (only date given 1840). It had a cast iron false plate with 'Richard Blood, Birmingham' on it. The style of the dial appeared to be about 1820, so he must have set up in business about that time, though so far this is the only recorded dial.

There were several firms which were associated with one another in various ways, so an attempt will be made to analyse them together, though one of them, William Finnemore II, goes up to 1852.

WALKER & FINNEMORE

Walker & Finnemore of Edmund Street were in business for only a short time (1808–11), but they produced a steady stream of high quality dials. The very earliest dials resembled Osborne or Wilson gesso work, such as this one (**Illustration No. 46**), made for John Bennett of Norwich, whose known dates of 1790–9 cannot be strictly accurate as he must have been working at least until 1808. The gesso work is elaborate, and the graphics very well done. The author has a theory that Walker and Finnemore were too very different personalities that probably did not get on all that well. This idea is only backed up by the dials themselves. Walker, the more

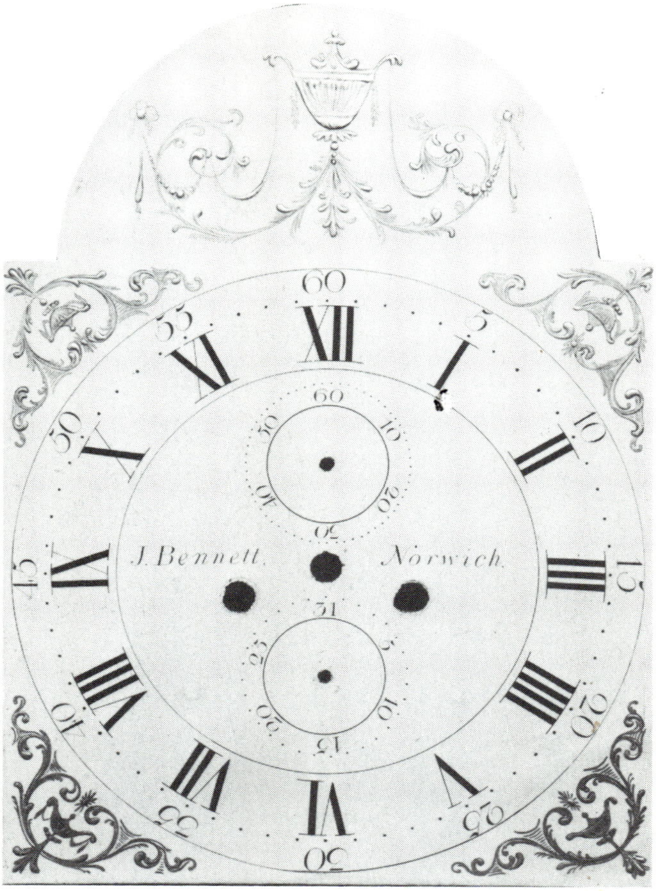

46 An early Walker and Finnemore of Birmingham dial made for John Bennet of Norwich *c.* 1808 with very fine gesso work in the old style, but there are urns in the centre of the corners which denote a more modern style. Urns had, however, been used in break arch gesso decoration from 1790. See page 85

difficult man, painted a number of dials with a lot of fine lines, often resembling rays, and he seemed to have been obsessed with repeating groups of circles or diamonds. This can be better seen when discussing dials made by Walker alone.

Finnemore himself was very fond of roses and his rose corners are easily identified such as this one (**Illustration No. 47**) made for William Broderick of Spalding, probably *c.* 1811. This dial is very unusual in having a number, 987, on the dial, which presumably meant that this clock was the 987th clock Broderick made. The numbering of clocks on the dial face is much more common in the United States. The minute markings are also interesting as they give an almost snowflake effect.

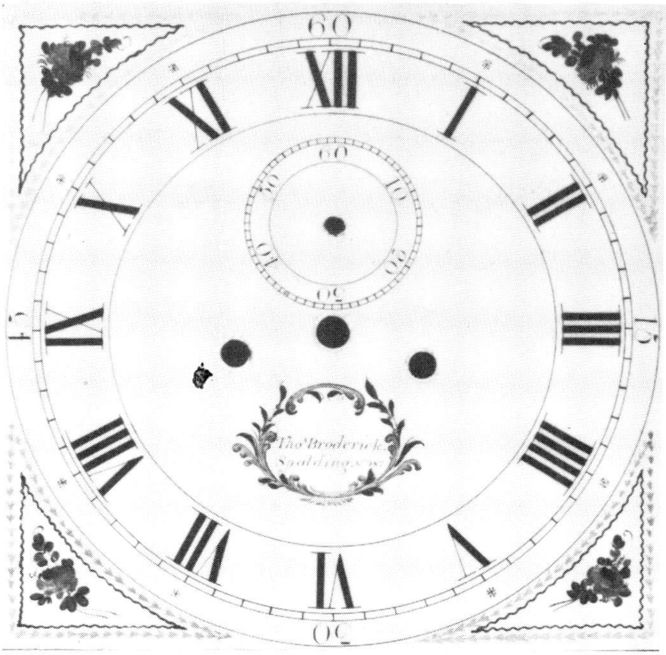

47 Walker and Finnemore with typical rose corners made for Thomas Broderick of Spalding *c.* 1811. It is unusual to have the number of the clock on the dial. See page 86

The graphics on Walker and Finnemore dials include full Romans and Arabics on some. Others had flared Arabic numerals with smaller Arabics on other quarters and the presumably later variety had Roman numerals and small Arabic quarters. The variety of graphics used during this short period of only three years or so shows just how eclectic the choice of graphics could be.

Walker and Finnemore produced dials with well painted pictorial corners. The subjects were the four seasons, the four continents and the four virtues, though other themes could have been used. One particularly interesting dial was made for the famous Samuel Harlow of Ashbourne, who wrote *The Clock Maker's Guide*. It has large single peaches together with two sprays of blackberries in the corners, straight tumbling Arabic numerals as well as a barred minute circle. None of this is really notable, but what is unusual is the moon dial where the moon face is identical to a Wilson one made for Stephen Hurst of Liverpool (1805–13). It is certain that after Wilson died in 1809, this particular artist went over to work for Walker and Finnemore. To the untrained eye, many moon faces look the same, but each artist has his own style, though this style may alter as time goes on. Moon faces painted at roughly the same time by a particular artist are usually identical. The scenic parts of the Harlow and Hurst moon dials are different, but the painting of the waves in the seascape is identical. None of the other Walker and Finnemore moon dials appear to have been painted by this artist. It has been suggested that the Walker and Finnemore Wilson type moon could have been a Wilson replacement, but this is doubtful as the moon dial was plainly stamped, which could not have been done after the dial was painted without damaging the paint.

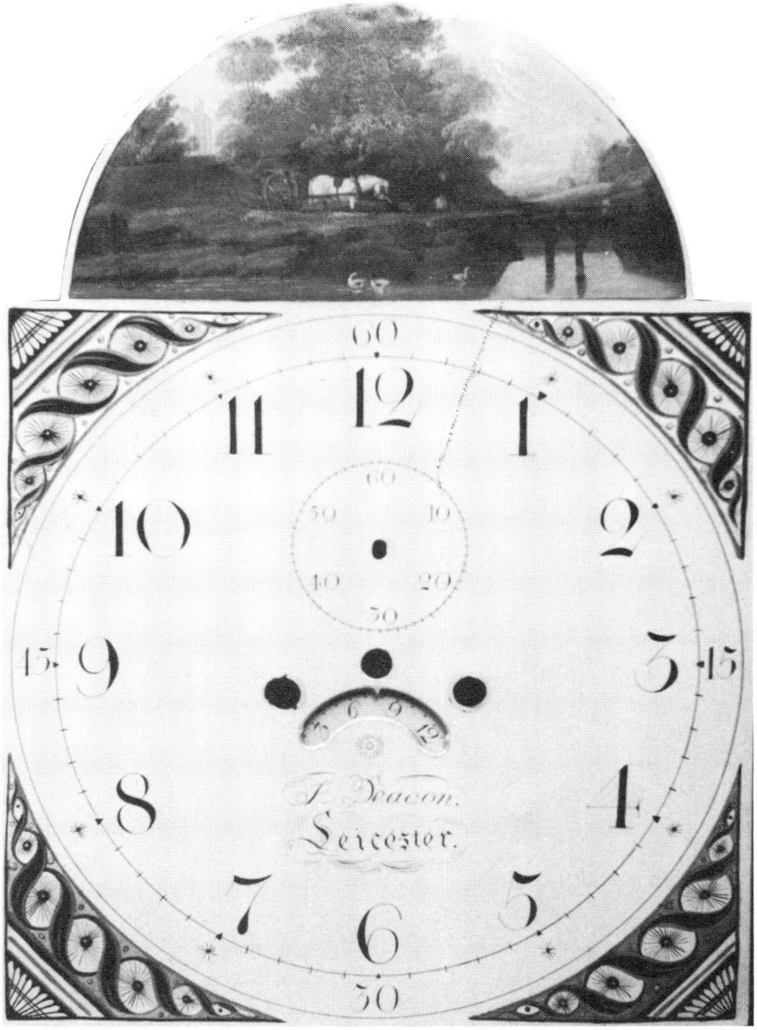

48 These corner motifs are typical Walker work and appear almost to move. The break arch is very well painted. The dial was made for John Deacon of Leicester. Walker worked alone for just one year during 1811. See pages 89–90

In 1811, Walker and Finnemore parted company. Each one carried on alone, though Walker worked alone for only one year. All of this can be seen in a most amusing way from the false plates. Finnemore filled in the Walker part of a master false plate and Walker did the same to the Finnemore on another master. It is not known just how many dials Walker produced. The author has restored only two with definite attribution but has obtained a good colour photograph of a third. The first one, an elegant oval, was made for Samuel Deacon of Barton, which will be discussed in detail in the later section concerning restoration.

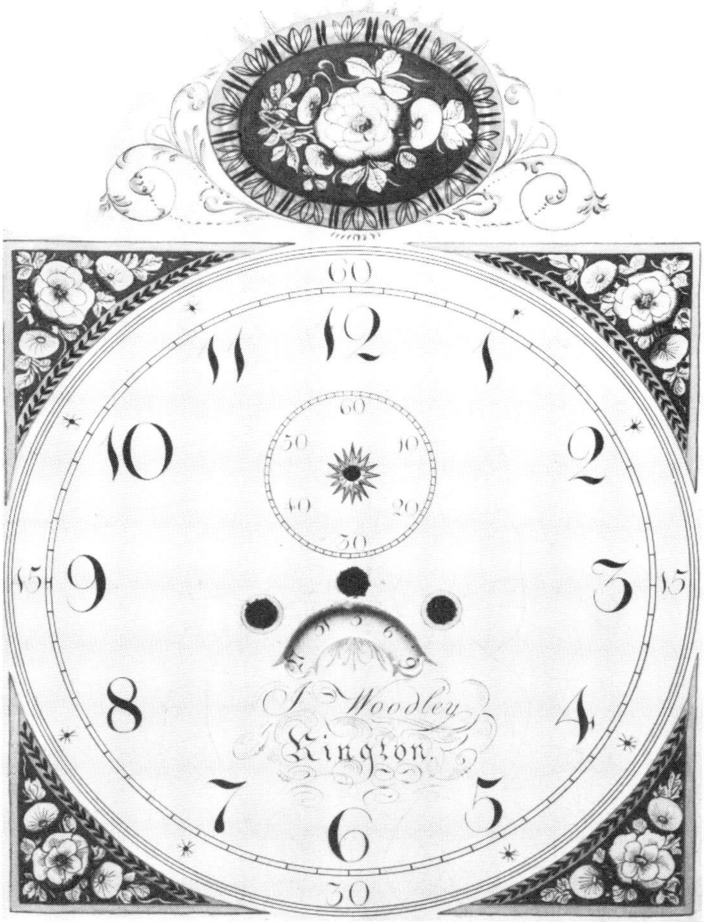

49 An unusual Walker dial using coppery bronze powder background to the corner and break arch roses which were very stylized. The scrolling around the name, Woodley of Kington, is over-elaborate and rather unbalanced. See page 90

The second dial (**Illustration No. 48**) arrived with 'James Price, Bosworth' overpainted on it. There was a fine Deacon movement, and the original name 'J. Deacon, Leicester' showed up clearly under the UV lamp. The strange corner motifs are typical Walker work. The third Walker dial is a real oddity. It was made for J. Plowman of Chichester, with Roman numerals and Arabic quarters, but the decoration in corners and break arch can only be called weird. The artwork consists of diamonds with holes in the centre in a circular pattern which appears to rotate, together with bare, rather menacing looking branches. Two doctors studied this dial and other Walker work, and both were of the opinion that Walker probably suffered from migraine or schizophrenia, or possibly both.

WALKER & CO

The next dial to be considered is rather a conundrum in many ways (**Illustration No. 49**). The cast iron false plate is Walker & Co, which neither Loomes or McKenna list. The name of the clock maker is equally elusive. A Josiah Woodley, who worked together with his son in Kineton, is listed by Loomes as 1828–50, which is obviously far too late. In the addendum to Loomes's Dictionary there is another Woodley of Kineton, *c.* 1770, also listed. There exist two Kinetons, one in Gloucestershire and the other in Warwickshire. There are three Kingtons, one in Gloucestershire, one in Herefordshire and one in Worcestershire. Kington is on the dial and it is the Worcestershire one that seems most likely as it is larger in size than the other two, and could easily have had a clock maker working there in the first part of the nineteenth century. These are all the facts that can be discovered about the clock maker, none of which is very conclusive.

However, the artwork and graphics tell a great deal more. The corner motifs consist of semi-double roses with a most unusual coppery bronze powder background paint. The roses are identical to those used in some Walker and Hughes dials where the corners have a background of brass filings set in glue. Therefore it can be presumed that the Walker of Walker & Co was the man who originally worked with Finnemore. Even more conclusive is the highly original scrolling with its unusual repetitive circles around the name of the town, Kington. Possibly the 'Co' could have been Hughes before the partnership was made definite. The author has gone into a great deal of detail concerning this dial in order to show what can be discovered when the dictionary and other sources seem to fail. It would be most interesting to know if more 'Walker & Co' dials turn up and, if so, what they are like. There must be others, for on p. 39 of Brian Loomes's *White Dial Clocks* there is a drawing of a Walker & Co false plate.

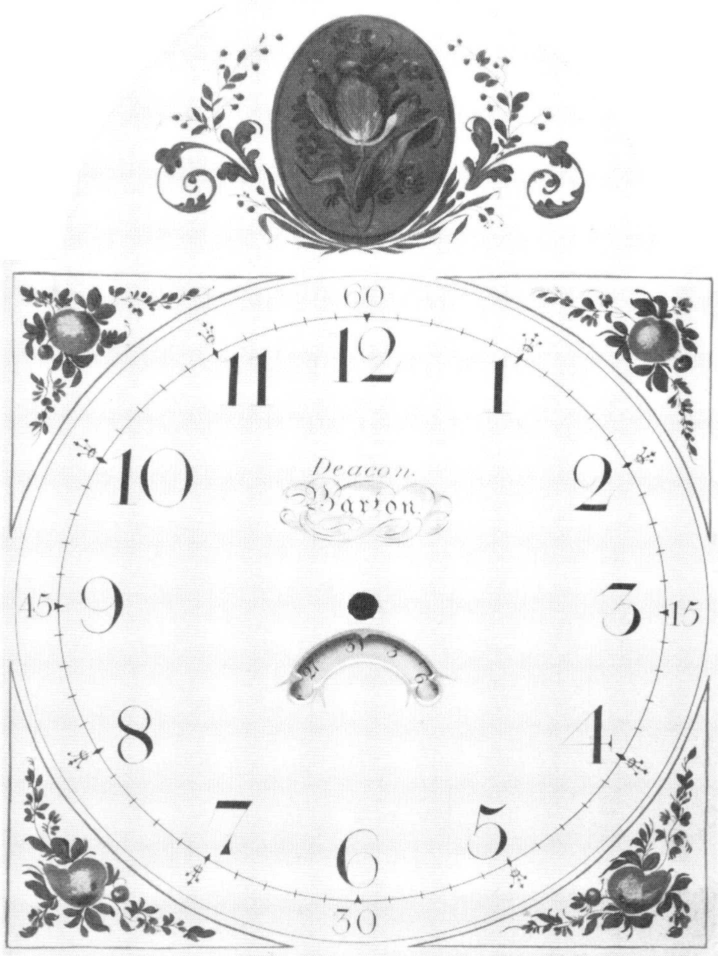

50 Walker and Hughes dial of 1812–13 with green background to the tulip in the break arch and the Walker and Hughes trident minute markings. It was made for Samuel Deacon of Barton. See page 92

WALKER & HUGHES

The Walker & Hughes partnership lasted a long time, from 1812–35. They worked at 15 Lower Priory, and later Fisher Street and produced a great number of dials, some of which were very fine indeed. It could be said that Walker and Hughes, together with the Finnemores, produced a large proportion of Birmingham dials in this middle period of 1812–35. As a matter of interest, a rather crudely made dial with a cast iron false

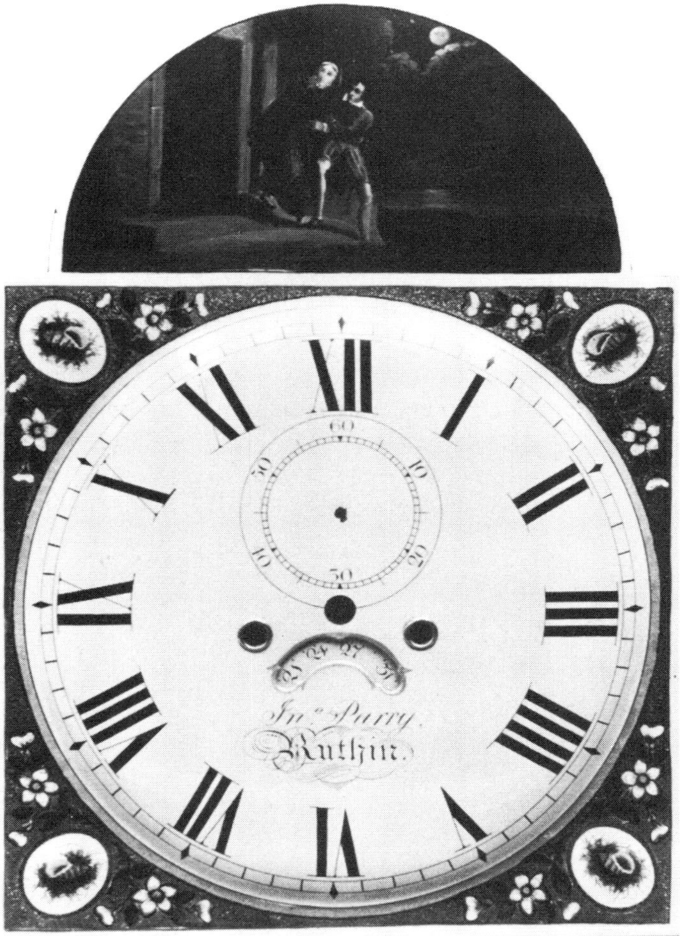

51 Walker and Hughes break arch dial made for John Parry of Ruthin *c.* 1830 has brass filings in the corners as a background which give a glittery appearance. There is a well-painted romantic scene in the break arch. See page 93

plate marked 'Hughes & Co. Birm' turned up for restoration. It has an irregular cut-out in the break arch with a rocking ship and flared tumbling Arabic numerals with quarters. It was made for F. Abraham of Yeovil, which is an unlisted name, but the dial appears to be no later than 1820. Whether or not this firm has anything to do with the Hughes of Walker and Hughes is not known, but the style of the dial does not resemble anything from the partnership.

The earliest dials made by Walker and Hughes were those with straight set Arabic numerals with Arabics at the quarters and interim trident minute markings. These tridents, whose prongs sometimes numbered more than three, appear to have been exclusive to Walker and Hughes, but other motifs were used as well. Another distinguishing characteristic was the rather narrow date aperture, though around 1825 some wider ones do appear. The earliest dials could have fruit and flowers enclosed in a lightly painted roughly triangular border such as this one (**Illustration No. 50**) made for Samuel Deacon of Barton (1812–13). The use of a parrot tulip in the break arch has been seen before in the same position though without the green background, as used by Osborne in a clock dial exported to the United States which can be viewed in the author's family home, 'The Bryant Homestead', Cummington, Massachusetts (Boston Trustees of Public Reservations). The triangular borders tended to get heavier as time went on and developed into a denticular or dog toothed variety, sometimes done in gold leaf or coloured paint, which appears not to have been used by any other maker. Sometimes this denticular border could be decorated by a number of fine lines. Shell motifs, usually with attendant seaweed, were often used but florals and geometrics also abounded. There were Wilson type birds and floral sprays in the centre of some dials as well.

The same type of rose as seen in the corners of the Walker and Co. dial was used either with a plain background or the brass filing one, though the flower could vary. They were always painted in the same style, using crimson lake for the details. The use of these brass filings was almost exclusive to Walker and Hughes. Though one firm, Wilkes & Son (1820–31), attempted to copy this style, it did not look the same at all. This dial (**Illustration No. 51**) was probably made for John Parry of Ruthin (1834–44) sometime in the 1830s. It is a good example of this type of corner, which also includes an oval with a shell. The flowers are of the single variety with the usual details in crimson lake, which invariably has faded to an undistinguished brown. This type of corner dates from 1815 to 1835.

Another characteristic of some Walker and Hughes dials was the very fine paintings done in the corners and break arch. The author had been working on such a dial (**Colour illustration (xii)**)with a scene in the break arch which looked familiar. When on holiday in Rome, she saw the same scene on a large tapestry in the Vatican made from a painting by Reubens, 'The Presentation of the Infant Jesus at the Temple'. There were obviously several good artists employed by Walker and Hughes, and one of them painted moon dials in a very recognizable style up to some time in the 1820s. The type of ship painted in the seascape varied, as did the expression on the moon face, but the basic way of painting the moon remained the same. The waves of the sea were always done in the same rather regular manner. There was often a large off-white Georgian house on the landscape side.

Walker and Hughes made some very unusual dials to special order. One such is this very large 16″ × 23″ dial made for John Winstanley of

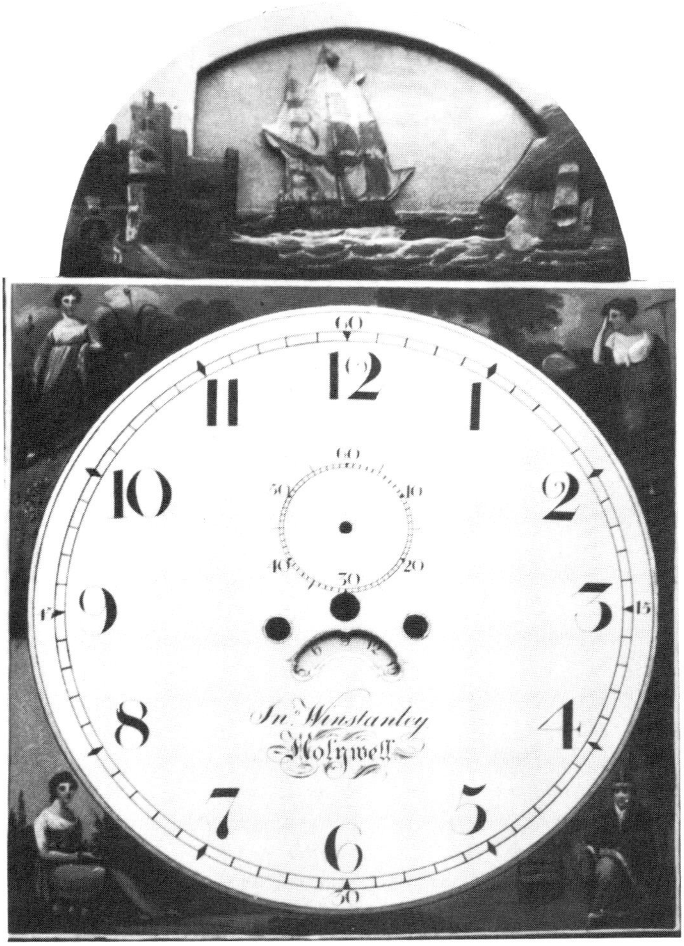

52 A very large special order Walker and Hughes dial made for John Winstanley of Holywell *c.* 1812. There is a rocking ship automata in the break arch with has an irregular cut out which is unusual in this period. The corners represent the four seasons. See pages 93 and 94

Holywell (1791–1840) about 1812 with a rocking ship in a battle at sea of some sort (**Illustration No. 52**). The very regular painting of the waves is identical to that found on the 'Georgian Houses' moon dial. Also, the irregular cut-out for the automata scene is unusual at this period. One interesting thing is that the four feet as originally placed on the dial had been sawn off and repositioned more widely apart in order to support such a large dial. This had caused some wrinkling of the paint where the feet had originally been, though of course this was done before the dial was painted. The corners have been well painted and represent the four seasons. The Arabic quarters have shrunk to a very small size indeed, and are obviously on their way out. Walker and Hughes seemed to specialize in these very tiny Arabic quarters. Another characteristic of the Walker and Hughes graphics was the very fine scrolling around the name, which can be seen in all of the three Walker and Hughes dials illustrated.

Another rarity was an 1820s dial with a break arch moon, but the aperture of the moon was a crescent cut out just under the numerals for the moon's age. This left a large area in the centre of the break arch, which was painted with a fine landscape. The moon faces were set in a starry sky and the corners had the sort of flowers usually associated with the brass filings. Roman numerals were used, together with tiny Arabic quarters, and the name John Warry of Bristol (1775–1830) had the usual very fine scrolling around it. The use of imitation jewels by this maker has been mentioned before, but it might be interesting to describe the two dials which featured these stones. One was made for John Smith of Wrexham (1777–1830). It had flared Arabic numerals, very small Arabic quarters with trident interim markings and two birds in the centre. The corners were geometric but dissimilar, that is, the top right and bottom left were one pattern, and the top left and bottom right were the other pattern. The break arch had a yellow ochre oval, into which the topaz-coloured stone was set, and this was flanked by two parrot tulips. The name had the usual fine scrolling around it. The second dial with stones is even more interesting. It was made for Ashmead of Newnham, dates unknown, and had flared Arabic numerals with small quarters and an Adam and Eve automata in the break arch (**Colour Illustration (xi)**). The corners featured circles of diamonds with a lot of small lines coming from them so as to give the appearance of movement, and the stone appeared in the middle of the circle. The corners are obviously typical Walker work.

Walker and Hughes also used gold leaf quite a lot in the dial corners. Sometimes it was used as a background to the design or in the L-shaped denticular borders. It was more rarely incorporated into the ladies' dresses of the four seasons where detail was added on top of the gold leaf. Another late development unique to Walker and Hughes from 1820 onwards was the use of strange flowers either in the dial centre or in the corners. The flowers looked like a yellow pansy on the top, with a red or blue morning glory underneath. The effect of this was rather curious and definitely heavy.

Another rather strange item came to light when the author was looking over a very good illustrated booklet, 'Hull and East Riding Clocks', by J.E.S. Walker, which served as a catalogue for an exhibition at Kingston upon Hull in 1982. On p. 53 was a dial that shouted Walker and Hughes, with a denticular border surrounding sea shells in the corners and flared Arabic numerals plus tiny quarters. Yet it had an Owen false plate. On the back of the dial, however, was a small paper label printed 'Manufactured by Walker and Hughes'. Both firms worked in Lower Priory, but Owen's dates (1803–05) preclude any co-operation between the two firms, so it is possible that Walker and Hughes used up some old Owen false plates. This does make identification rather difficult, as small paper labels tend to get lost.

The use of bronze powder in the unusual Walker and Co dial has been mentioned, and it was used again about 1820 in a dial made for William Underhill of Newport (1828–36) where it formed the background for all of the Roman numerals in the chapter ring as well as featuring in the shell corners and the break arch. The author considers that first Walker and then Walker and Hughes were probably the first dial makers to regularly use this bronze paint. Later on other makers took up this type of paint and used it a great deal, especially as backgrounds for scenes or figures. Another feature unique to Walker and Hughes was the use of the 'Bearded Lady' in no less than three dials in the 1810–20 period (see **Colour Illustration (xiii)**). The artist painted ladies very well in dials representing the four seasons, but for some reason the lower right figure representing Winter was supposed to be a man. The profile is identical to the ladies, but with a beard on it.

Walker and Hughes can be considered to have been a very important firm in the history of dial making, not just because of the quantity and quality of the dials they produced, but for their original and innovative ideas. Luckily, most of their dials had the firm's name on them, either cast into the false plate or stamped on the date and/or the moon dials. They certainly did export dials to the United States, as evidenced by one made for John Holt of Lancaster, Pennsylvania, in about 1815. This dial has shells in the corners with an L-shaped denticular gold leaf border, straight set Arabic numerals, tiny quarters and centre date markings. An illustration of this dial can be seen on p. 120 of *Clockmakers of Lancaster County and Their Clocks, 1750–1850* by Stacy Wood Jr and Stephen E. Kramer III.

FINNEMORE

Meanwhile, the other part of the original Walker and Finnemore founded the firm Finnemore, that continued on from 1812–52. They occupied the premises at various times in numbers 3 and 4 Edmund Street, Calthorpe Street and Hospital Street. It is probable that the Finnemores produced as many dials as Walker and Hughes, but many of the later ones were unmarked. There were four different Finnemores but the cast iron false plates had either just plain 'Finnemore' on them or 'Finnemore and Son', though at first the 'Walker and Finnemore' master was used with the Walker partially obliterated. There follows a list of the Finnemores, but it is impossible to differentiate between their work.

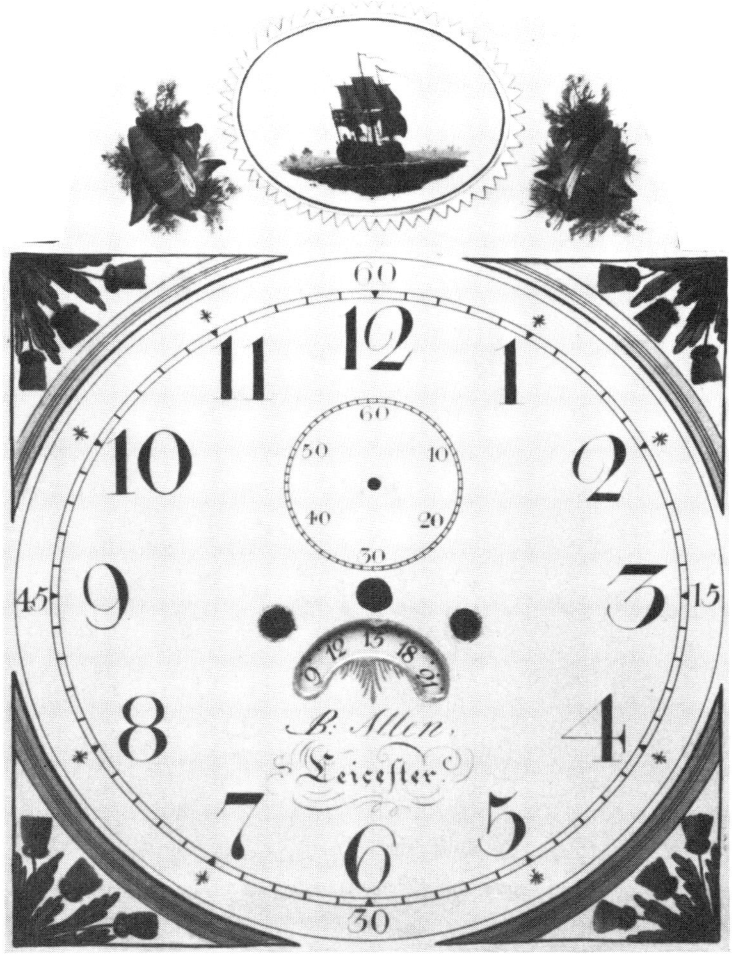

53 Finnemore of Birmingham dial *c.* 1815 with B. Allen on the dial, though another undecipherable name had been underneath. There is a typical Finnemore ship in the break arch and the flowers in the corners are meant to be thistles. See page 98

William I	1812–25
William & Son	1828–36
William & George	1839–46
William II	1849–52

The early 1812–20 Finnemore dials with Arabic numerals were very similar to the Walker and Hughes ones, though there seem to have been a preponderance of dials with roses in the corners. Long-tailed birds

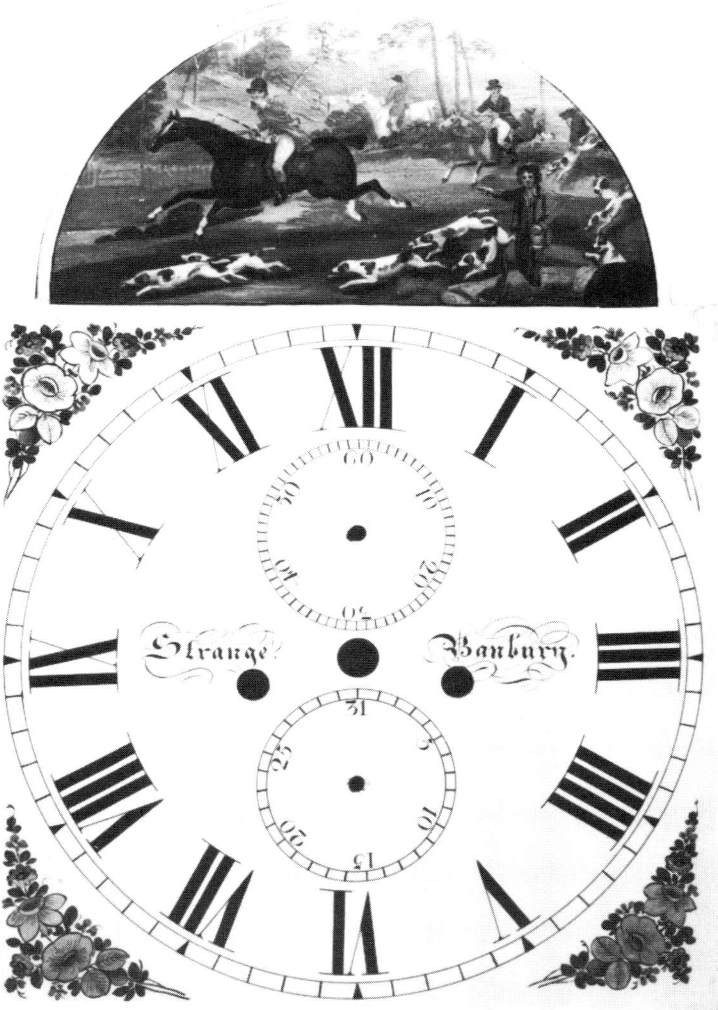

54 Typical Finnemore gold roses in the corners of this dial made for Thomas Strange of Banbury *c.* 1830. There is a well painted hunting scene in the break arch. It is unusual that no grommets were ever fitted. See page 99–100

appeared in the centres as well as flowers, but compared to Walker and Hughes there seemed to be a lack of really fine art work, and Finnemore dials tended to be a bit heavy both with regard to graphics and decorative work. Of course after 1825 most dial makers' products tended to be a bit ponderous anyway. One element often found on Finnemore dials was a ship, usually painted in the break arch such as this dial (**Illustration No. 53**) made *c.* 1815 for B. Allen of Leicester. The ship could be quite

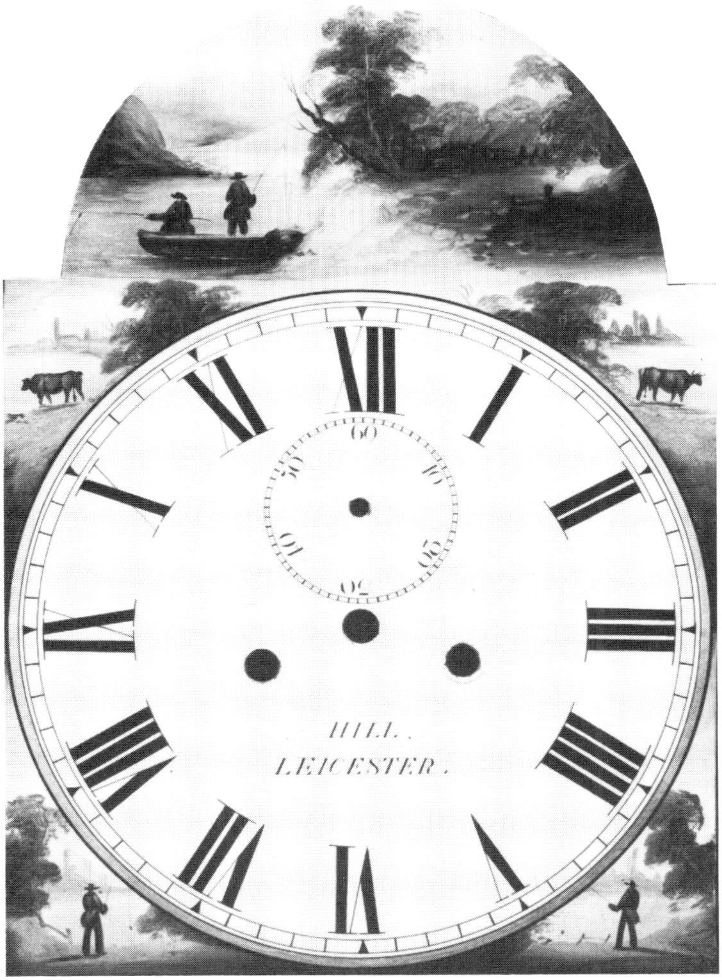

55 A late (1842–52) Finnemore dial made for Hill of Leicester which features the blue-clad fishermen that are seen on some other Finnemore dials of this period. The false plate, which is original, was made from an old Finnemore dial *c.* 1815–20. See page 100

large, surrounded by sea, though sometimes in rather worrying proximity to the land. The scrolling on Finnemore dials was not so elegant as that on the Walker and Hughes ones. As well as ships and roses, the blue convolvulus or morning glory appeared many times, but the gold leaf flower as seen in **Illustration No. 54** also was used a lot. This dial with its hunting scene in the break arch was made for Thomas Strange of Banbury *c.* 1830. These hunting scenes were very popular from the 1820s to 1850s, and this particular one appears to be quite well painted. The dial never had the brass grommets fitted, which is unusual for the period. Another decorative feature sometimes seen in the break arch of Finnemore dials was the large off-white Georgian house done with an astonishing type of perspective that makes the house appear to be sliding downhill.

Sometimes it is possible to trace a design back quite a way in the firm's history. This is true of the star-pointed, half-oval corner with darts first used by Walker and Finnemore on a dial made for Thomas Bartley of Denbigh (1781–1811). (See **Illustration No. 134, 135**) The half oval was done in gold leaf with interior geometric decorations in pink and the rest of the corner featured yellow ochre. The darts fanned outwards from the star points. The same corner was used in a Finnemore dial made for Rees Williams of Cwmmusk (sic) (*c.* 1815–20) and another one in a different colour scheme using red was made for Robert Williams of Wrexham (1842–44) which also had the name in a Wilson-like ribbon.

One variation of the flower corner appeared in the Finnemore and Son period of 1828–36. The background in the corner was black, often with a small gold leaf motif in the apex of the triangle and boldly painted roses, usually pink, which gave quite a dramatic effect. As far as is known, Finnemore was the only maker to use black in this way, though plenty of dials by other makers had green, navy or Prussian blue backgrounds to the flowers in the corners.

Another type of very late Finnemore dial, probably belonging to the last period (1849–52) was the 'Blue-clad Fishermen' style made for Hill of Leicester (**Illustration No. 55**). One interesting thing about this dial, apart from the subject matter, was that the false plate, which was original to the dial, had been made from an earlier (1815–20) Finnemore dial which included an oval in the corner. These ovals were a favourite motif in Finnemore dials of that period.

Another distinctive dial of the 'Blue-clad Fishermen' type had a royal blue background to the gold flowered corners as well as being all around the oval break arch scene. In addition, there were rudimentary scrollings on either side of the arch scene. The dial had been made for C. Collingborne of Brinkworth, and the false plate was made of a trial piece for a one-piece silvered dial complete with marking out and engraving.

This type of gold scrolling has turned up in the corners of several more dials, usually in conjunction with flowers, such as this one (**Illustration No. 56**) made for either Hugh Roberts Sr. or his son Hugh Jr, probably in the late 1840s. This dial had an unmarked sheet iron false plate. The break arch scenes at this period were quite often rather crudely painted.

The style of the flower painting was quite distinctive, and these flowers have turned up on slightly later dials, minus the gold leaf

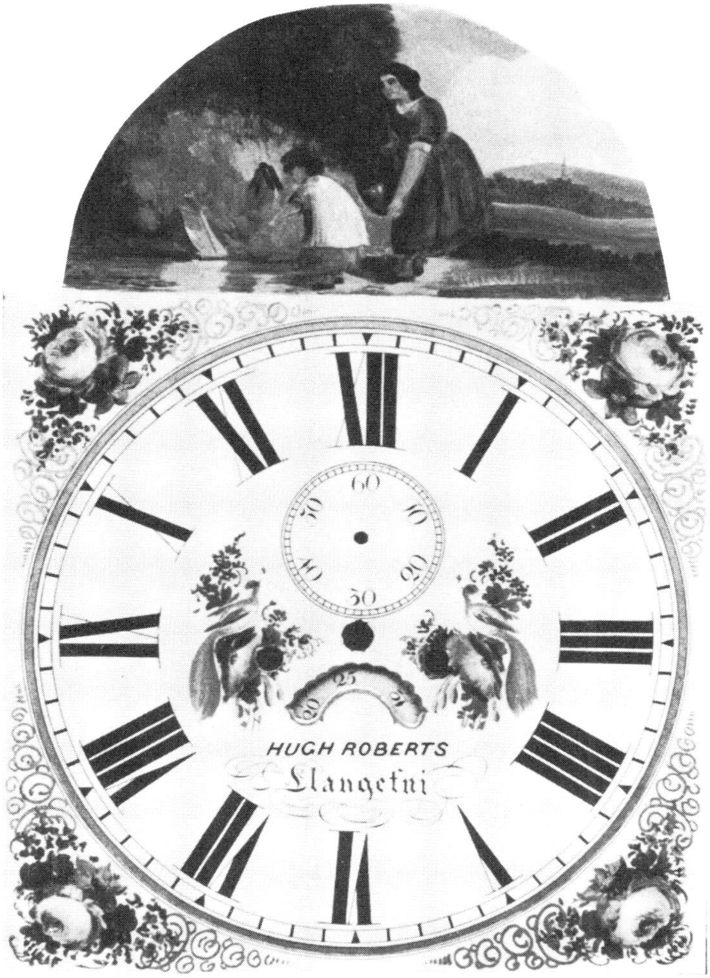

56 A typical late (1842–52) Finnemore production with overblown pink roses and gold scrolling in the corners together with the rather distinctive long-tailed birds in the centre. The break arch painting is thin and rather carelessly done. This dial was made for Hugh Roberts of Llangefni and had an unmarked sheet iron false plate. See page 100

scrolling but on a background of gold-coloured bronze paint. There are two examples of this type of dial, one of which is on p. 258 of Brian Loomes's *Grandfather Clocks and their Cases*. This particular dial has two peacocks boldly painted with roses in the break arch. Another, later dial is on p. 129 of Brian Loomes's *White Dial Clocks*. It has a scene with camels and palm trees in the break arch, also with a bronze paint background. It is important to realize that a progression of this sort, beginning with just one named dial, can only be made where accurate photographic records are kept.

Another interesting variation of the late Finnemore dials is one with castles in the four corners, birds and roses in the centre and in the break arch the automata of a man on a horse with a background, intending to portray the Battle of Waterloo. It was made for R.B. Thomas of Bank Place, Port Madoc, whose dates are given as 1868–95. Unfortunately, the horseman, presumably Wellington, was missing, so a replacement had to be made. The cut-out for the automata included trees, a cannon and even one cannon ball.

Having dealt with some of the firms which were quite well known in the period 1808–52 as well as those which were relatively obscure, the author will continue to examine the products of other dial makers, whose output ranged from very few to quite a considerable number. Because of the overlapping in time of the various firms, it is impossible to survey all the dial makers chronologically, and some firms, of course, seem to have either very vague dates or no dates at all. A firm not often encountered is Hobson and Hodgkinson, whose dates of 1815 or later, as given by Loomes, are not very exact. The author has restored only two dials by this maker. One had flared tumbling Arabic numerals and quarters with a sort of leaf interim marking, birds in the centre and flowers in an L-shaped corner border. It was made for John Smith of Wrexham, *c.* 1815. The second one had Roman numerals with Arabic quarters, the same leaf interim markings, shells in the centre and geometric corners. It was made for William Owen of Llanrwst (1810–35).

E. HOBSON & E. HOBSON & SON

It is quite possible that the Hobson of Hobson and Hodgkinson became the firm of E. Hobson and E. Hobson & Son, whose dates are given by Loomes as *c.* 1830, though it seems likely they were working earlier than this as evidenced by one dial made by E. Hobson and Son for Samuel and Lazarus of Carmarthen (dates rather vague) with straight Arabic numerals and quarters, shell corners and two ships in a battle scene in the break arch. The same artist painted a full seascape with one large ship in the break arch of another slightly later dial which had Roman numerals only which was made for Thomas Edwards of Corwen (1808–43). Both of these dials had a cast iron false plate with 'E. Hobson and Son' on it.

Hobson dials were usually well painted, and one particular dial is quite spectacular (**Illustration No. 57**) in its decoration, which leaves only the chapter ring plain. It was made for John Jones of Caernarvon (Carnarvon on dial) (1828–44), probably *c.* 1830. These fully painted centres were popular off and on for a long period, but they were

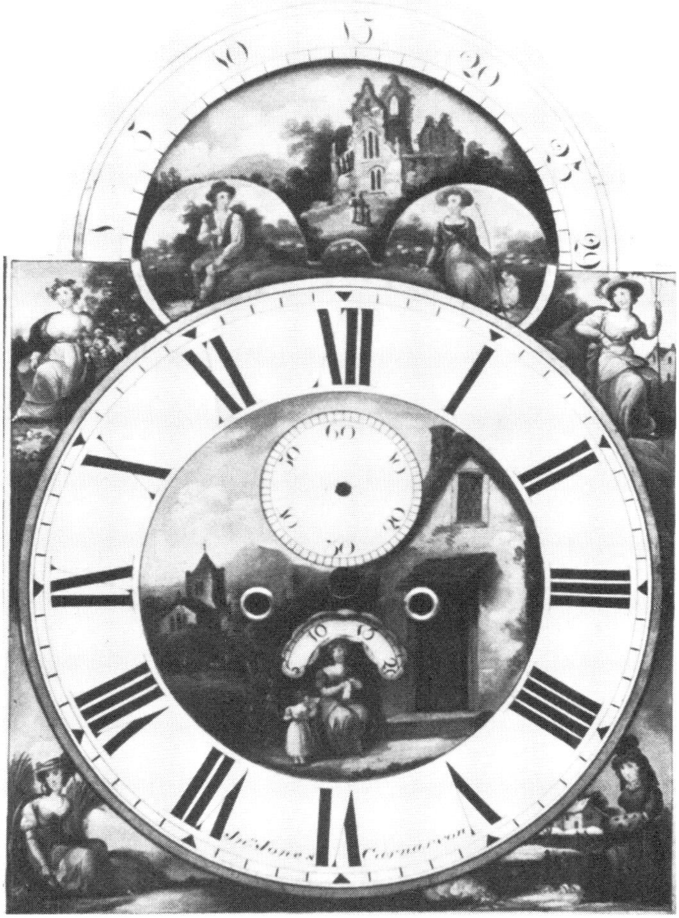

57 Hobson of Birmingham made this fully painted dial in about 1830 for John Jones of Carnarvon. The standard of the artwork is very high. See page 102

obviously more expensive. Another E. Hobson & Son centre painted dial turned up in Australia where the figures of mother and child are almost identical and the corners, though slightly different, are obviously painted by the same artist. Hobson also produced a very unusual dial for the same maker. It was a rather fat oval with a plain chapter ring and plain centre located in the lower two-thirds of the oval, a 12 o'clock moon dial with vague worlds on the hemispheres, two pretty girls in a landscape above this, and on the top part of the dial a temple scene with irregular cut-outs and a moving Father Time against a background of pale blue sky. One can only call a dial of this type a unique tour de force.

Another speciality of Hobson dials was the use of brilliantly coloured parrot tulips with small red flowers painted above them, which were usually found in the dial centre. The author has restored two of this sort of dial. One, made for Thomas Heywood of Bangor (1828–35) had a romantic looking edifice which was a combination of cottage and castle in the break arch, Roman numerals and rather pensive looking sheep with lambs in the corners. The second dial was made for John Winstanley of Holywell (1791–1840), also with Roman numerals, but this one had bright pink and blue geometric corners as well as a moon dial in the break arch and maps on the hemispheres (**Colour Illustration (xv)**). There must be quite a few unmarked Hobson dials of the 30-hour variety that have gone unrecognized because this firm painted the corners with a wide variety of motifs and unless the 30-hour dial matches up to one that is known to be a Hobson dial, correct attribution is impossible.

WILKES & BAKER

Wilkes & Baker (1815–20) of 10 Whittall Street is another firm whose products are not seen very often. The author has restored only three dials from this maker, and all were made for clock makers in Wales. The finest was a dial with tumbling Arabics and quarters, snowflake interim markings, baskets of flowers in the corners, and a moon dial featuring a windmill. This dial had John Parry of Tremadoc (1835–40) on it, but underneath and undecipherable was another presumably earlier name. The two other dials, both square ones, are suspected repaints with rather amateurish looking flowers in the corners, painted in a style which uses oil paints as if they were watercolours. These repaints were done a very long time ago by dial makers wanting to update the dial, probably *c.* 1840–50. The dial had 'R. Thomas Carnarvon' on it, done without serifs, and Roman numerals with late style full Arabic minute numerals. Robert Thomas worked from 1778–1827, so presumably his name was carried over on to the repaint. The second dial has Roman numerals only, but a square date aperture with a date ring, and is marked 'W. Owen, Llanwrst' (sic) in a slightly earlier style of graphics. William Owen of Llanrwst's dates are 1810–35, but any time after 1800 a square date aperture and date ring would be unusual.

A John Wilkes is listed as working in Hospital Street from 1803 until 1817, though the author has had nothing marked Wilkes from this period. John Wilkes then, obviously, joined up with Baker to become the Wilkes and Baker previously mentioned. In 1820 Wilkes and his son started up business at 10 Whittall Street, where the former partnership had been. The cast iron false plates of Wilkes & Son went on being used up to 1831, though Samuel Wilkes the son seems to have worked independently at 13 Whittall Street from 1829–51, when he moved to Great Hampton Row for

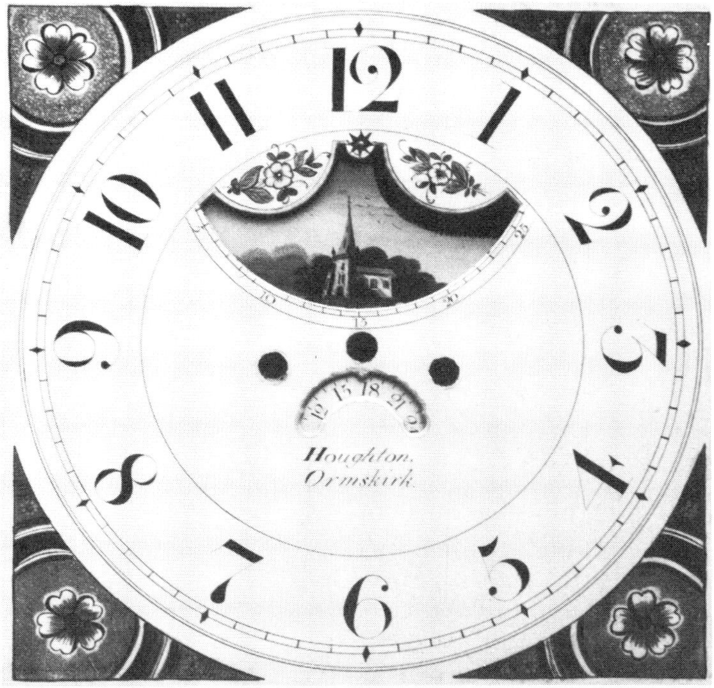

58 Wilkes and Son of Birmingham have attempted here to copy the Walker and Hughes brass filing corners, but the effect is altogether different. It was made for Houghton of Ormskirk *c.* 1820. See pages 104 and 105

three years. A very interesting cast iron false plate 'S. Wilkes, Whittall Street, Birmingham' turned up, unfortunately without a dial, and this false plate seems unique in having the address on it.

Wilkes & Son, of 10 Whittall Street (1820–53), made an effort to copy the Walker and Hughes brass filing corners in this example (**Illustration No. 58**) made for Houghton of Ormskirk *c.* 1820. It is the only one of this type that the author has seen, and it is instantly recognizable as not being the same as the Walker and Hughes dials. For example, the depth of the coating of brass filings is far thinner, giving a flatter effect, and the single flower is quite rudimentary. There are quite a number of Wilkes dials about, though only the earliest (*c.* 1820) still used all Arabic numerals. The corner decorations were varied, from shells, fruit and flowers through geometrics, four seasons, four continents, etc. and landscapes.

Wilkes made quite a few dished or convex dials, but one speciality as far as corners went was the rather plain type with a partially gold leafed flower with foliage which was outlined in dark blue or green. These must have had the great advantage of being very quick to do, and thus

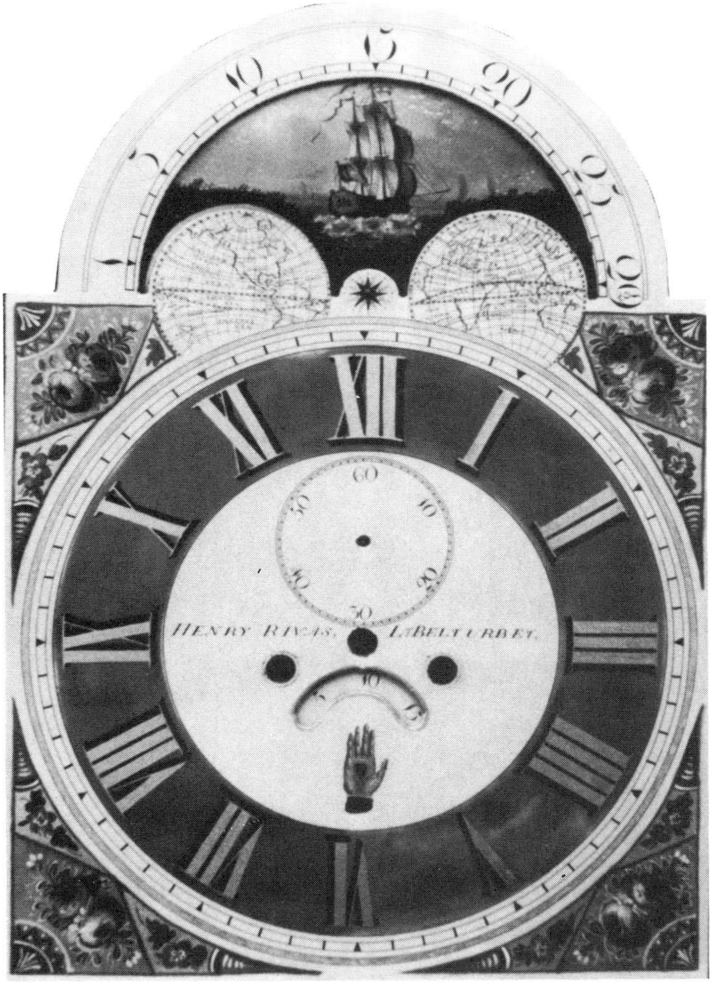

59 A very unusual Wilkes dial with a bright blue chapter ring and gold Roman numerals outlined in black. It was made for Henry Rivas of Little Belturbet, Ireland *c.* 1840. See page 106

cheaper to produce. Wilkes used both maps and vague worlds and figures or scenes in the hemispheres, but one interesting thing is that he changed his map type around 1840 to the sort with outside latitude markings. Most of the Wilkes break arch moon dials have wide arches which contain the numerals for the moon's age, which tends to give rather a heavy appearance. Another innovation that came in around the 1840s was the use of the bronze powder 'gold' background to the figures of the four seasons or four apostles and others. Wilkes certainly used this gold effect, but so did other makers. A really startling idea was the use of a bright blue chapter

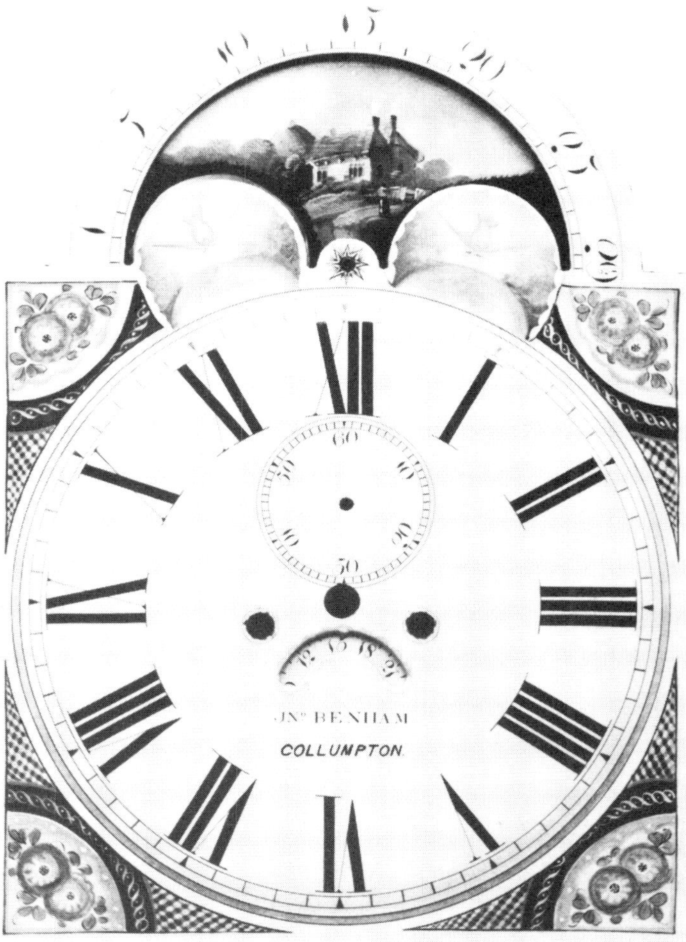

60 This is a dished or convex dial made by Wright of Birmingham for John Benham of Collumpton *c.* 1835. The maps of the world have been superseded by vague impressionistic worlds with the dove of peace. See page 108

ring with gold leaf numerals outlined in black (**Illustration No. 59**) at about the same 1840 period. This particular dial was exported to Ireland, but no trace of the clock maker Henry Rivas can be found. Wilkes used the older style maps on this dial. The painting of the heart in the hand could have been added in Ireland. Wilkes used both cast iron and the newer sheet iron for false plates, and in the case of the sheet iron false plates these were stamped with his name.

CHRISTOPHER WRIGHT

Christopher Wright (1835–45) worked at numbers 130 and 386 Summer Lane. His products were not particularly distinguished in any way, but his use of scenic effects was quite attractive, and he sometimes used very brilliant colours, especially in the four seasons dials. He also made dished dials such as this one (**Illustration No. 60**) which features what appear to be auriculas in the corners. It was made for John Benham of Collumpton *c.* 1835, and in spite of the fact that this dial uses the doves in the hemispheres, Wright continued to use the older style maps on some of his break arch dials. Somewhere in this period a new style of graphics appeared as far as the name or town was concerned. Note that the Benham name is in one style and Collumpton is in a completely different 'sans serifs' style. This 'sans serifs' type of graphics was far easier to do, and thus it appears on a great many later dials.

JOHN WRIGHT FLETCHER

Aside from William Finnemore II, John Wright Fletcher (1841–55) seems to be the last of the Birmingham dial makers who marked his work. The author does not know if the Wright in his name was anything to do with Christopher Wright, but both men worked at 386 Summer Lane. Christopher Wright was there for all of his productive years (1835–45) and John Wright Fletcher worked there from 1849–51. He started off in 1841 at 124 Great Hampton Street and in 1852 went to 25 Barr Street for one year, moving from there to 173 Hockley Hill from 1854–55. Fletcher appeared also to have made a lot of dials with scenic effects. These always had a white border area around the decoration, so that it did not appear to 'run off the dial', but this type of painting is far harder to do than when one can continue painting right up to the edge. Late dials of other makers tended to be painted in this easier way as speed in production became all important.

A tremendous number of dials exist of the 1840–70 period where there are no marks of any dial maker at all, so they can be classified only by type. In this late period it was not considered important to mark dials, as it took more time. Usually the later the dial the poorer the quality of preparation, artwork and graphics, though there are quite a few exceptions. Although much information has been given in 'Analysis of Style Changes 1800–70', it might be useful to comment in a bit more detail about this period.

SCENIC DIALS

Scenic effects, in square, break arch and those dials with a break arch moon, were popular right up to 1870. They were done with varying degrees of expertise, and some left a plain border edge unpainted, though the majority were painted right up to the edge of the dial as has been mentioned before. These scenic dials often contained cottages, castles, churches or ruined abbeys. One interesting thing to note was the millstone (just a circle with a dot in the middle) which seemed to be used in some dials as the focal point. Sometimes there were two millstones. Many of these scenic dials were unnamed, though some are easily recognized by the subject matter, such as Dunfermline Abbey or the harbour at Haverfordwest. One variation of the scenic dial occurred when the background to the building was done in either bronze powder or silver/tin leaf (see **Colour Illustrations (xvii) and (xviii)**). This could include the automaton of a rocking ship in the break arch, seen through an irregular cut-out representing rocky grottoes. Some of the buildings painted were romantic figments of the artist's imagination and probably never existed. Often people spend a great deal of time trying to identify these quite impossible constructions.

FOUR SEASONS DIALS

Four seasons dials, usually represented by female figures (although sometimes by boys), remained popular up to the end of the painted dial. Often in the late period the bronze powder background was used. As the years went by the fashions changed, and one stunning dial made for J. Camponovo of Wolverhampton (untraceable) with a gold background included ladies some of whose dresses were medieval in style. Where there are two or more dials done in the same way and decorated by the same artist, it is probable that more exist. There are three very similar dials, all with four seasons corners, made for the Huntingtons of Chester (*c.* 1840–50). One has an unmarked cast iron false plate, and the rest have an unmarked sheet iron false plate. As seen in **Illustration No. 61**, the artwork is distinctive, including that done on the moon dials. One of these Huntington dials had, unfortunately, to be stripped and repainted. (This will be dealt with in the section on restoration.) As far as classification goes, these dials should be called 'Huntington Dials' until any further evidence turns up as to their real origins.

The four continents also continued to be used, and although the figure of Britannia remained essentially the same, some of the others varied considerably. One very large break arch moon type dial, over 16″ wide, which was made for the Morath brothers of Liverpool (1848–51), featured an 'American' Indian with curly hair, moustache and goatee beard, poling himself along in some sort of canoe, while Africa was

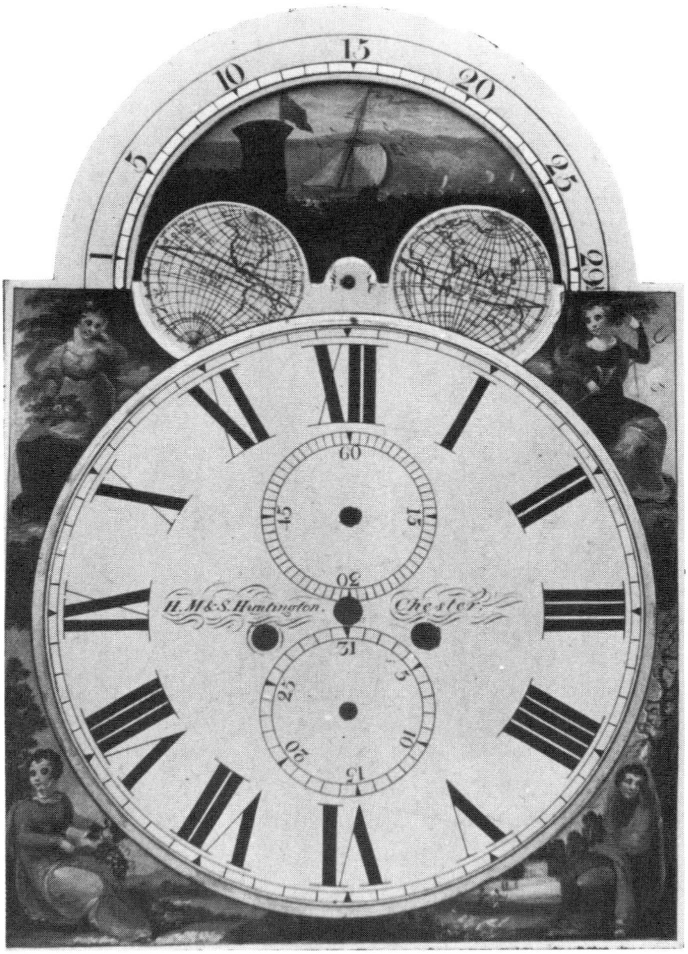

61 One of three similar four seasons dials all made for the Huntingdons of Chester *c.* 1840–50 by an unknown Birmingham dial maker. See pages 109 and 195

represented by a black girl decently clad in a white chemise. This dial had a centre date, and in the large middle area David played his harp to King Saul. A seconds dial was set into this, and the landscape backgrounds to the two top figures continued into the hemispheres **(Colour Illustration No. (xxiii)**. The author has been assured that another identical dial exists, but thus far it has not been seen.

FLORAL DIALS

Floral dials were used up to 1870 and were painted in a great variety of ways. One type of floral dial is quite distinctive, and at least six examples

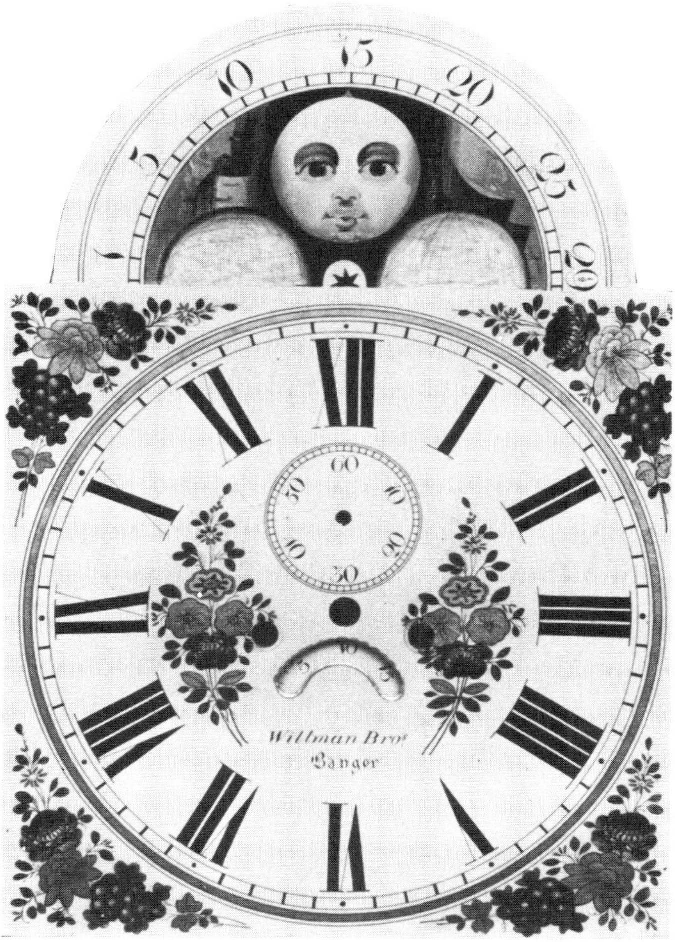

62 A good example of the painted-on false moon made by an unknown Birmingham dial maker for the Willman Brothers of Bangor 1856–60. This dial is in the 'gold flowers' style on a cream-coloured base. See pages 110 and 111

of this variety exist. All are on a cream base with rather heavy Roman numerals and a date aperture with two bunches of flowers either side of the aperture. The flowers always include some done in gold leaf, though the same variety was not always used. The style of painting is distinctive. Four of these dials were square, one had a fully painted break arch scene of a mother and child at a well and the last is a glorious example (**Illustration No. 62**) of the painted-on moon, which was meant to convince the onlooker that the clock had real moon works. The clock makers for whom these dials were made worked from 1850 onwards, and the William brothers, Joseph and Sebastian, for whom this dial was made worked in Bangor from 1856–87. All of these dials suffered from some degree of what appears to be paint incompatibility, and the base paint was very thin. Also the varnish had been put on very thickly with no attempt to follow the outline of the flowers.

It is true that many dial makers, including the Finnemores and Walker and Hughes, used gold flowers, but when as many as six dials exhibit exactly the same characteristics, they should be classed as a type.

There are, of course, many thousands of fine dials where the dial maker is unknown, and a great number of these seem to come from the

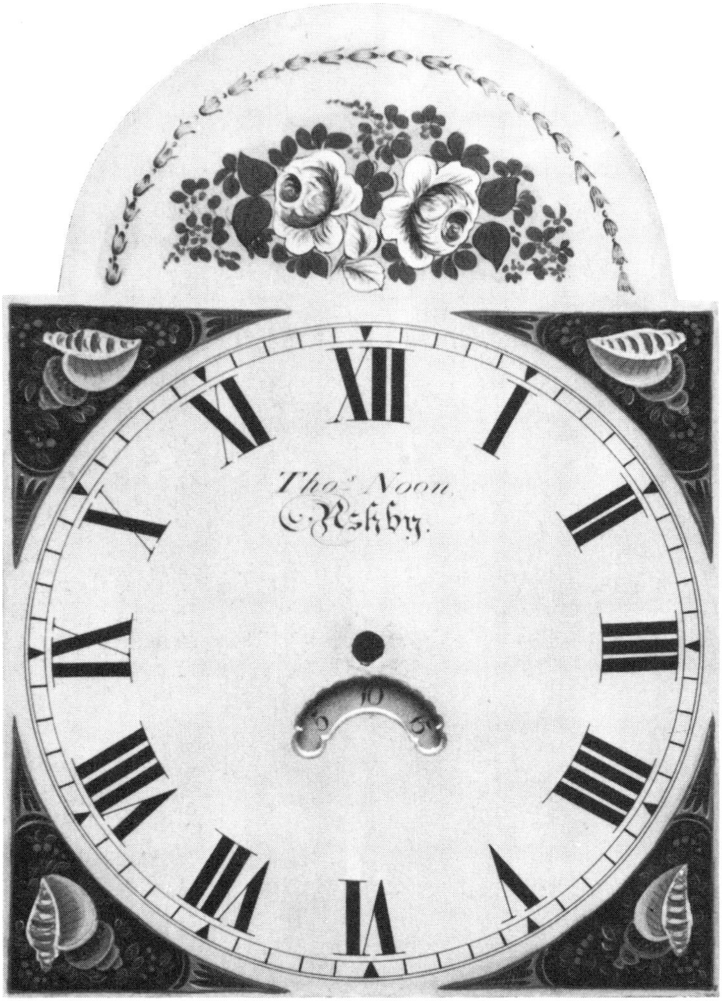

63 Unknown Birmingham dial maker made this attractive 30-hour dial for Thomas Noon of Ashby *c.* 1835. See page 113

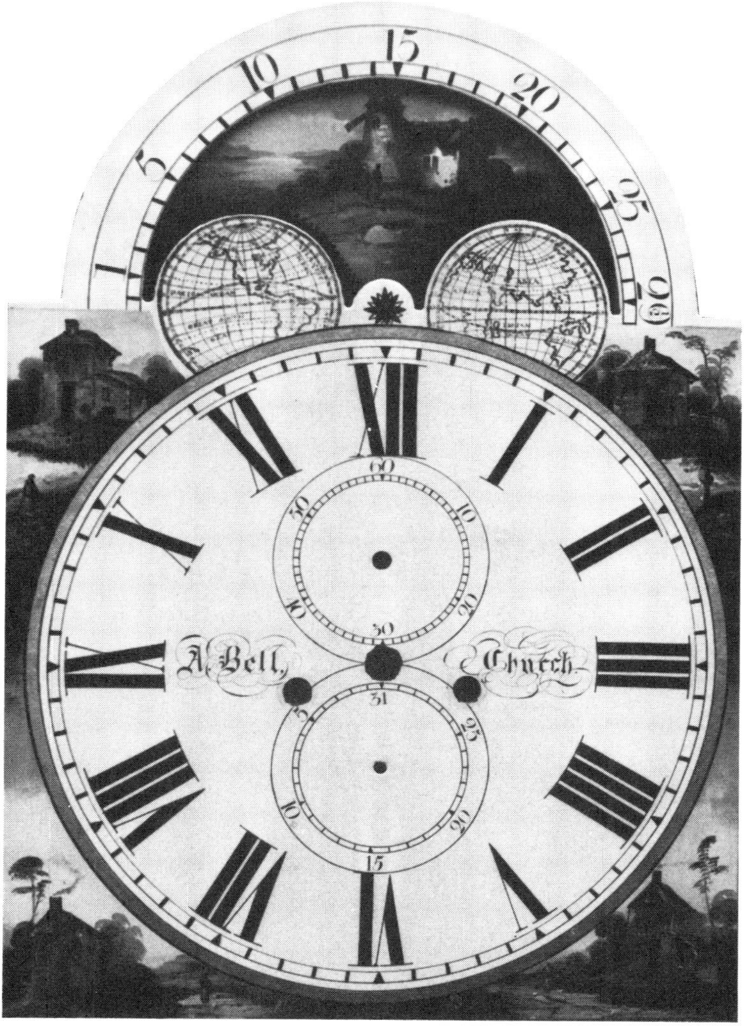

64 A very different style of Birmingham-made dial from an unknown maker, where silvery leaf has been used as a background to the cottages in the corners. There exist two dials of this type, both made for A. Bell of Church *c.* 1840. Note the maps have outside lattitude markings. See pages 113–14

later period after 1830. A lot of them are of the 30-hour type, and this dial (**Illustration No. 63**) is a good example. It was made for Thomas Noon of Ashby (*c.* 1835) and used gold leaf for the roses in the break arch as well as in the shells for which the background was a navy blue. The case for this clock would have been relatively simple. Other dials are obviously more expensive and this one (**Illustration No. 64**) was made for A. Bell of Church sometime in the 1840s. The corners have silvery backgrounds, and the maps are of the late variety with outside latitude markings. The moon was made of iron, and this dial had a sheet iron false plate. Very often these late dials used iron for the moon, but one dial of roughly the same type as the Bell dial used zinc, which seems a poor choice as that metal is not very rigid and tends to shed paint.

UNUSUAL GRAPHICS

The statement has been made that the fully numbered minutes were not used after 1810, but there exist at least two original dials with this type of

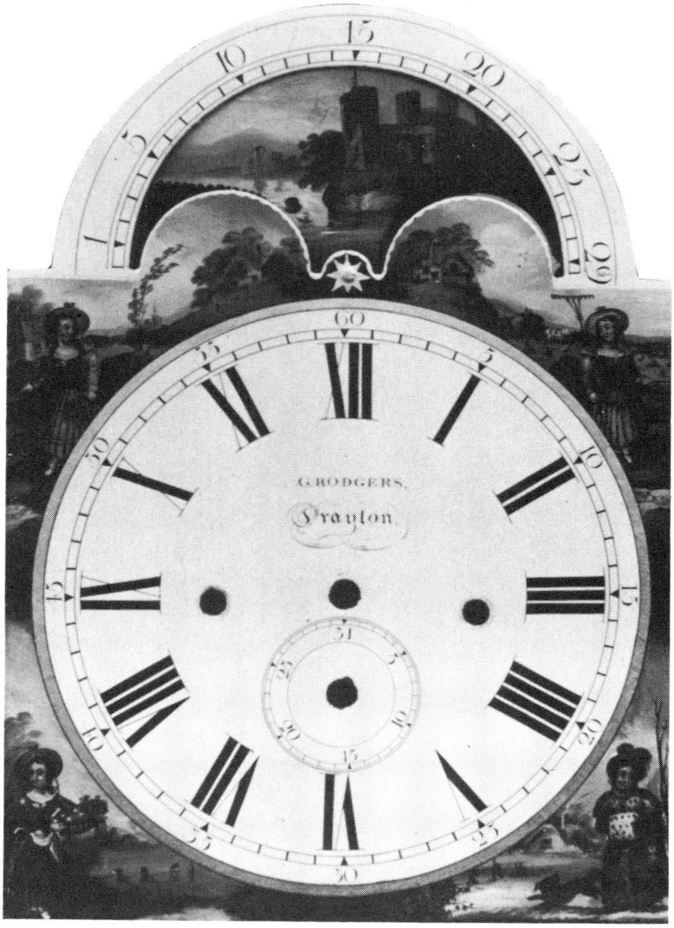

65 The silvery leaf background which was used in this four seasons dial is very unusual for the period (*c.* 1845) in that it has full minute numerals. It was made by an unknown Birmingham dial maker for George Rodgers of Market Drayton. See pages 114–15

graphics which definitely belong in the later period. Suspected repaints, however early, are not included. The first (**Illustration No. 65**) was made for George Rodgers of Market Drayton (1842–70). It used silvery leaf as a background to the four seasons corners, and in its original form included a date ring on the dial. The clock maker altered this, widening the hole to provide a winding hole for a musical train which, unfortunately, is now missing. The moon dial is unusual in that instead of the regular sailing ship there is a steam vessel, although equipped with sails as was common at the beginning of the steam age.

The second dial showing full minute notation is one made for J.G. Munro of South Shields (*c.* 1820). It has a centre date hand and is of the scenic variety with a named view 'Abbey Bisham' in the break arch.

These unusually late dials with full minute numerals can easily be recognized because the Arabic numerals are fatter than those elegant examples of an earlier period. Other types of usual graphics do exist, such as the 24-hour dial, the Franklyn type and the world time notation, but there do not seem to be very many of each of these varieties and, because they are often rather complicated clocks, the dials tend to be quite individual.

REPAINTED DIALS

A certain amount of confusion can exist about dating a dial where the name on the false plate does not correspond with the style of the painting or graphics, and sometimes the clock maker's dates do not match up with the dial maker. However, a word of caution here concerning the dates in various dictionaries, regional books, etc. These dates are taken from directories and census forms of the period, but they should be considered a guide and not absolute gospel. Both Wilson and Osborne dials were repainted, and one suspects Wilkes and Baker as well. A repaint where the decoration and graphics were never made by that dial maker is easy to spot, such as a scenic break arch and corners painted up to the edge of the dial with heavy Roman numerals, yet the false plate had 'Osbornes Manufactury Birmingham' on it. Those repaints done closer to the dates of the original manufacturer are much harder to recognize. However, in examining the suspected repaint in the context of other dials made by that maker, the repaint will stick out dramatically. Usually it is the decorative elements that show up here.

CONCLUSION

This ends the section on Birmingham dial makers, both known and unknown. The influence of the Birmingham makers was enormous, and their products seem to have been distributed worldwide. A great deal of work remains to be done, especially on the later dial makers whose work was often anonymous, but at least a start has been made.

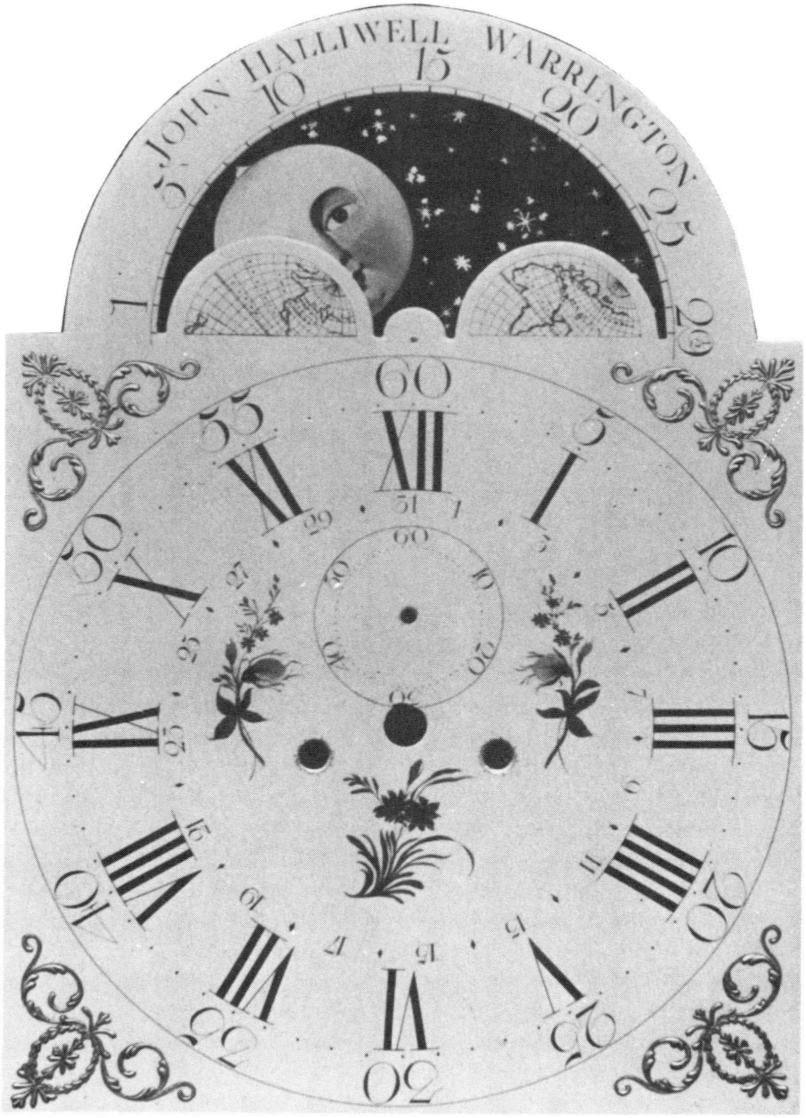

66 A John Halliwell of Warrington dial that he made for himself *c.* 1780. This is an exceptionally fine dial with gilded gesso work and gilded winding holes as well as most unusual snowflake-like stars in the deep blue sky on the moon dial. See page 117

4 Dial Makers in Places Other than Birmingham

EARLY UNKNOWNS

This section deals with the various other dial makers around the United Kingdom. None of these produced anything like the quantities that Birmingham made, but it is important to record whatever is known about these firms as it is all part of the picture. Often these smaller dial makers are of great local interest.

JOHN HALLIWELL OF WARRINGTON

John Halliwell's dates are given as 1790–1820, but it is very probable that he worked prior to 1790. He was a clock maker as well as a dial maker. Two dials exist which he made for himself. The first one (**Illustration No. 66**) has an unmarked cast iron false plate and the moon dial is distinguished by some very snowflake-like gold leaf stars. The restrained gesso work is of a high quality and the maps, though similar to very early Wilson ones, are just a bit different. The winding holes are gilded, and the whole dial does appear to be pre-1790, as does the second one (**Illustration No. 67**), which has a cast iron false plate with 'I*H' on it. I and J were interchangeable at this period. The dial shows the distinctive flower paintings that were quite different from those of Birmingham. The gesso borders are similar to some used by Osborne. The date aperture is of the early square ended type, and the winding holes are gilded.

There must have been more Halliwell dials made, and there is a distinct possibility that he made an especially fine dial (**Illustration No. 68**) for Enoch Hawksey of Nantwich (1779–99). This dial appears to be slightly later than the other two, because brass grommets were used on the winding holes. The gesso work is the same as used on the square dial, but the cast iron false plate has no markings. Another possible Halliwell dial was made for Samuel Collier of Eccles (1750–1800) (**Illustration No. 69**). The gilded gesso corner framing

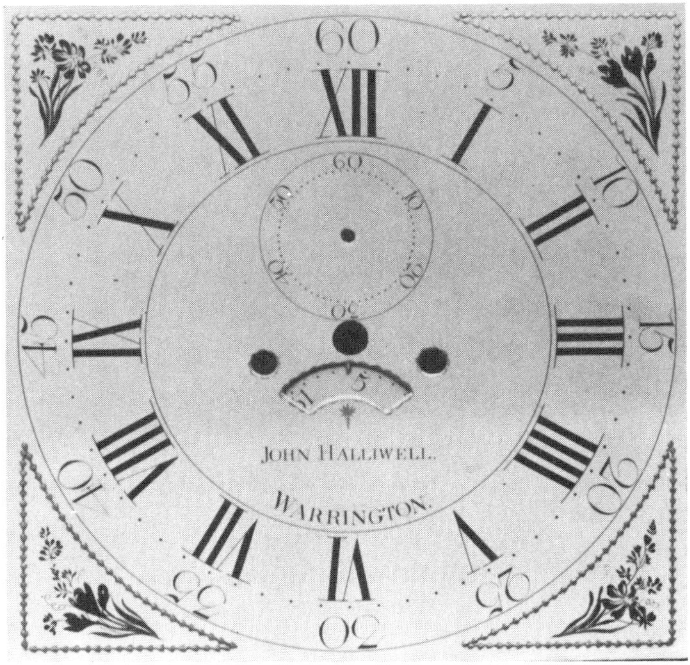

67 Another John Halliwell of Warrington dial from the early 1780s with a square-ended date aperture and gilded winding holes. The cast iron false plate was marked I*H. See page 117

has been replaced by red and blue paint in the same pattern as in **Illustrations Nos. 67** and **68**. The fruit and flower corners are similar to the Hawksey dial, and the date aperture is rounded. Once more the cast iron false plate is unmarked, which is also true of the date dial.

It is an interesting question as to what became of the John Halliwell dial-making firm after about 1790, which appears to be the date of the latest dial probably attributed to him. There were others of the same Halliwell name who worked north of Manchester: William at Chorley (1770–95), another William of Blackburn (1795) and James at Chorley (1820), though whether these were all related is impossible to tell. Possibly dial making did not pay John Halliwell as well as making clocks and watches, but the dials which remain are certainly outstanding productions.

MANCHESTER DIALS

Manchester or thereabouts certainly was a centre for dial making in the period 1780–1805 or later. The name of any individual firm is, unfortunately, not known. These dials are easily recognizable though at

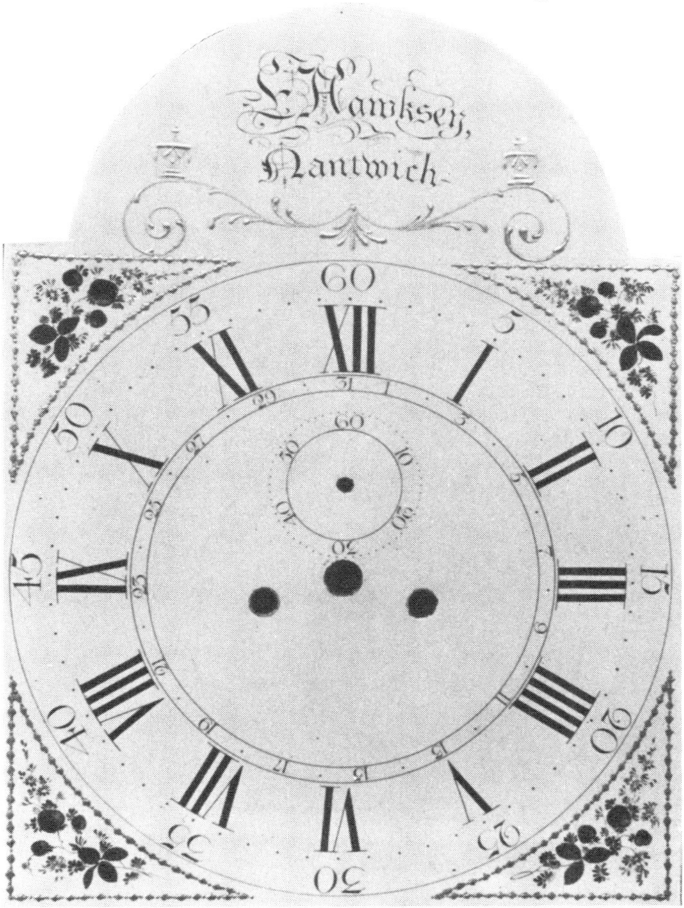

68 This is probably a Halliwell dial made for Enoch Hawksey of Nantwich *c.* 1785. It has brass grommets on the winding holes and an unmarked cast iron false plate. See page 117

least two different artists' styles can be seen, and the dials usually had four long feet. The gesso work, when used, is flatter than that of Birmingham or Warrington. The fruit and flower corners are also slightly bolder than those of Birmingham and Warrington and the birds are quite unique. One of the main features in those dials with moons is the long-nosed moon face which has a slightly mournful expression. The maps are also unique and were probably hand drawn, as no two are exactly alike. One unusual treatment of the area where maps were normally found was on a dial made for Major Schofield of Salford (1707–83), where there is a rather startling rising sun on the left hand side, and what could be a rudimentary map on the right.

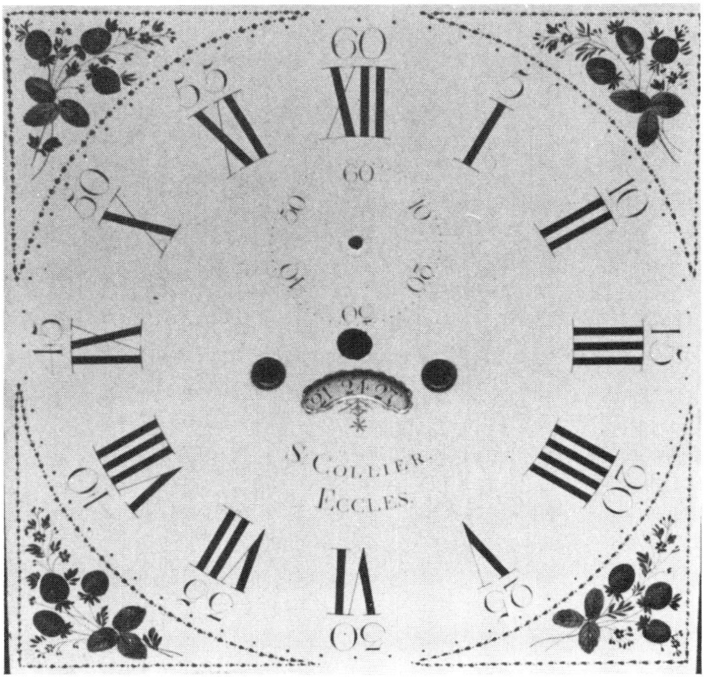

69 Although no markings are on the cast iron false plate, this dial looks like one made by Halliwell of Warrington around 1785. By this time date apertures were rounded. The dial was made for Samuel Collier of Eccles. See pages 117–18

About the earliest known Manchester dials restored by the author appears to be this one (**Illustration No.70**) made for Samuel Clare of Halton, near Runcorn *c.* 1780. It has a square ended date aperture, and the large date dial shows definite hammer marks on the back. The date dial was attached to the main dial by fitting the centre hole over a short stub foot which was drilled to take a pin. This method of attaching date dials was used by Osborne and Wilson in the 1770s, but was soon superseded in Birmingham by the more practical pipe and post method. The short stub foot method remained in use for nearly all the Manchester dials. Another characteristic of these Manchester dials was the fine art work. Popular prints often seemed to have been the inspiration for this, so it is not surprising to find two remarkably similar lots of corner paintings done on two different dials. The first one (**Colour Illustration No. (xix)**) is from a three train bracket clock made for Barker of Warrington and Wigan (date uncertain). It is based on brass, which is very unusual, and it has the usual Manchester gilded winding holes. The moon is not of the long nosed variety, and the maps are very rudimentary. The corners represent the four seasons and probably represent the four stages of love or a similar theme. The

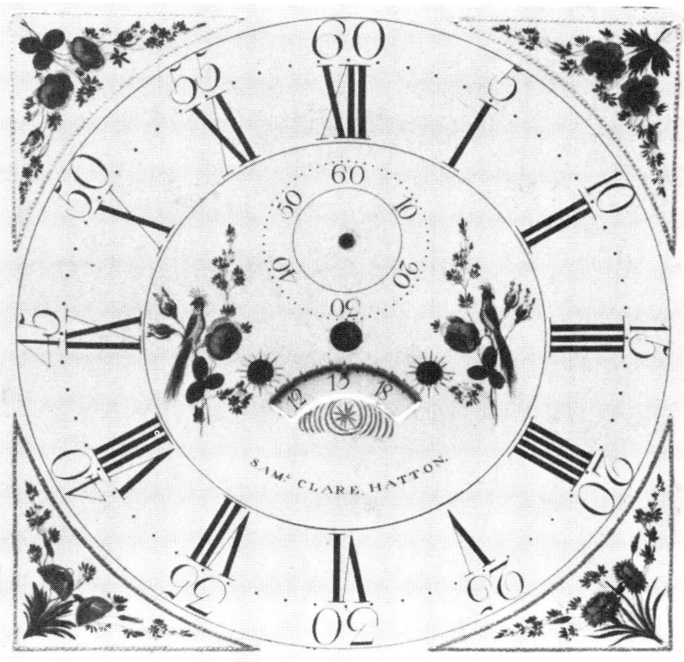

70 An early *c.* 1780 Manchester-made dial with a square-ended date aperture and a very large date dial as well as gilded winding holes. The birds are typical Manchester style and it was made for Samuel Clare of Hatton. See page 120

second dial (**Colour Illustration No. (xx)**) was made for Edward Shepley of Manchester (1788–1800). It has a most attractive centre painting with its rose coloured sky, and the bottom corners are very similar to those on the Barker dial. The curious maps can be clearly seen, and this dial does have a long nosed moon. The rose coloured background was used in the corners of another dial of the four seasons type made for Samuel Collier of Eccles *c.* 1790, which also had a long nosed moon face. Details of this dial can be seen on pp. 202–5 in the section on restoration.

A slightly different type of ground paint was used on at least one batch of Manchester dials *c.* 1795–1800. The paint, which was softer and greyer, has deteriorated, producing a multitude of very fine cracks all over. Two dials of this type exist, though there are probably more. The earlier one was a square dial with the usual type of fruit and flowers in the corners, typical birds in the centre, and a 12 o'clock moon which had a cheerful countenance. It was made for William Kirk of Stockport (1760–95). The second one was of the same type but with a large round ended date aperture, the date dial revolving on a short stub foot. The aperture was heavily decorated, and the dial was made for John Collier of Cheadle *c.* 1800.

As in Birmingham, the closer to 1800, the larger became the clocks, and the dials became correspondingly larger as well. A very fine 14½″ wide dial (**Illustration No. 71**), owned by Manchester University, was made for Edward Shepley of Manchester *c.* 1800. The long nosed moon can clearly be seen, and the curious maps are the most detailed ones from Manchester that the author has ever seen. A great number of rivers fancifully decorate the right hand maps. An almost identical dial was discovered in Alexandria, Virginia, close to Washington DC. It had the same type of long nosed moon, and maps, though the maps were in rather poor condi-

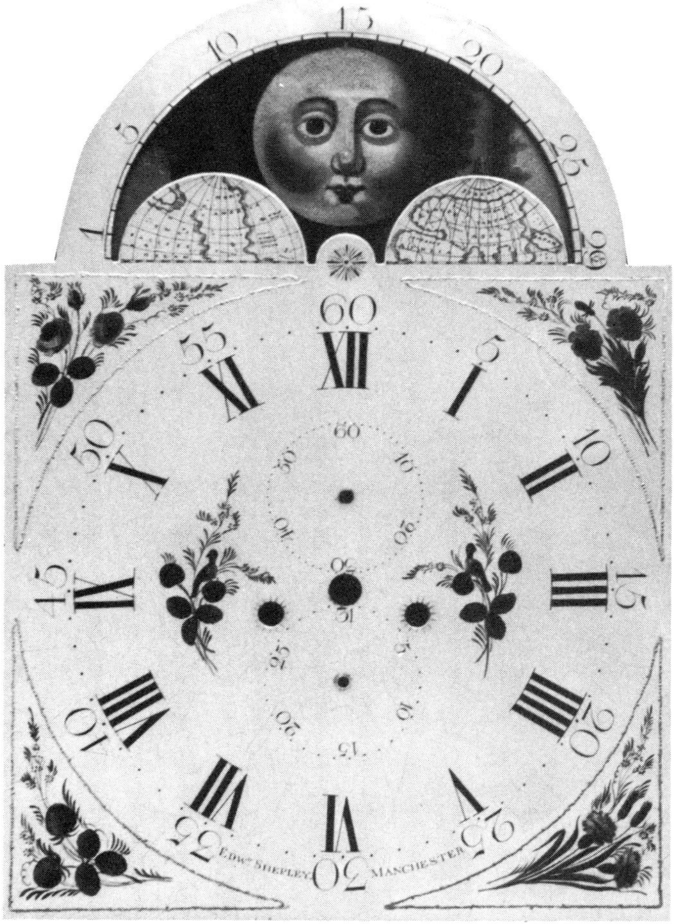

71 A large wide fine Manchester-made dial with very fine detailed maps of the Manchester type and a long-nosed mournful moon. It was made for Edward Shepley of Manchester *c.* 1800.
See page 122

tion. It had been made for John Kent of Manchester (1769–1813).

A very fully painted soft ground Manchester dial with religious connotations, was spotted, luckily, by a collector who realized that it probably was unique (**Colour Illustration No. (xxi)**). It was made for J. Buckley of Ashton, presumably John Buckley of Ashton under Lyne, whose only date is given as *c.* 1820. It had Roman and Arabic numerals with no grommets at all.

Occasionally it is possible to roughly date a dial by the subject matter, and such a dial was made in Manchester for Martin Clayton of that city (1789–1824). The centre painting depicted a memorial to Nelson, who died in 1805, so obviously this firm was still working in 1805 or later. The corners of the Clayton dial were of the four seasons type, but the style of dress has been updated. The maps remain substantially the same. After 1805 no other later Manchester dials have turned up, but possibly they do exist. One named firm of dial makers, Hardie and Christie of 17 Oak Street, Manchester, was known to have worked in 1824, but it is not known what sort of dials they produced.

HALIFAX AND NEWCASTLE DIAL MAKERS

Another separate style of dial making emerged in the Halifax–Newcastle area around 1790 and continued well into the 1840s. There were several firms involved.

Ralph Beilby, who had been associated with the famous engraver Thomas Bewick, joined up with his brass founder brother-in-law, James Hawthorn, to become the firm of Beilby and Hawthorn of Newcastle (1797–1817). They seemed to have formed a partnership with William Whitaker of Halifax, who sometimes worked on his own, but at other times Whitaker was associated with William Shreve, a painter. There were many varieties of cast iron false plates used by these firms, and it is almost impossible to give exact dates for each type of false plate. Some examples that have been noted are given here, with rough dating:

1795–1815	Top of false plate	Beilby Hawthorn N. Castle
	Bottom of false plate	W.W. Halifax
	(unusual bowed plate)	
	Top of false plate	Whitaker & Shreve Halifax
	Bottom of false plate	Hawthorn Newcastle
1803–1815	Top of false plate	Willm. Whitaker
	Bottom of false plate	Halifax
1800–1817	Top of false plate	Whitaker & Shreve
	Bottom of false plate	Halifax

It is also rather difficult to sort out the confusion between the various types of Beilby Hawthorn and Whitaker–Shreve dials, but some classification can be made according to the design of the artwork as well as the graphics. Also, a great number of these dials had large numerals done very boldly on the reverse of the dial, moon dial as well as on the false plate, though so far it has been impossible to correlate the numbers on the dials with any dates. One particular dial (**Illustration No. 72**)

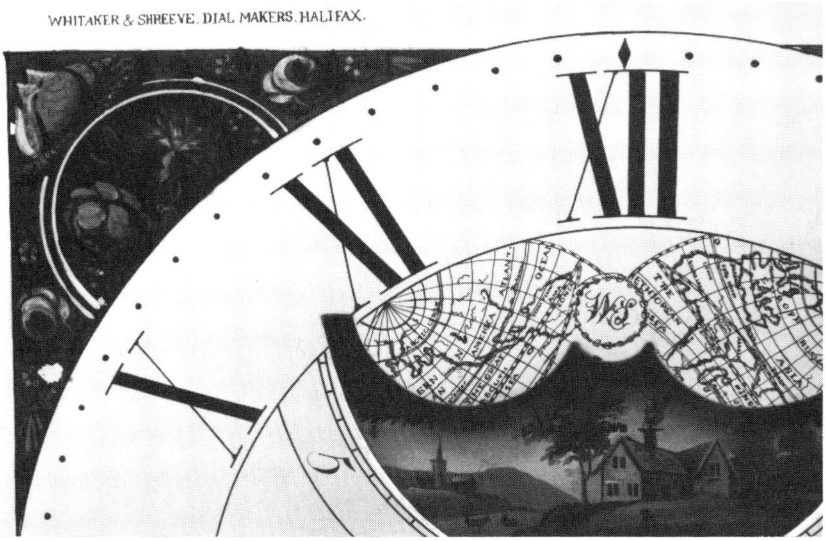

72 There is a rare transfer print on the corner of this dial showing the maker's name as well as intertwined initials between the maps. See page 124

made for Martlew of Ormskirk *c.* 1815 had a most unusual transfer print on the top left hand border of the dial: 'Whitaker and Shreeve Dial makers Halifax', and the intertwined initials W. & S. also transfer printed between the maps. One other dial with the print on the border is known to exist, but there well may have been more where the transfer print has rubbed off.

An early *c.* 1791 dial was made by the Whitaker and Shreve Halifax–Hawthorn Newcastle firm for John and William Blaylock of Longtown (**Illustration No. 73**). It is of superb quality and the false plate has an unusual bowed shape. This type of plain dial does not seem to have been made by this firm very often. An interesting detail is the graphics on 'Longtown', where the lettering is not solidly filled in but barred. This type of lettering has been found on a later dial from the Halifax Newcastle firms.

There are quite a series of dials depicting a young girl in a pink dress,

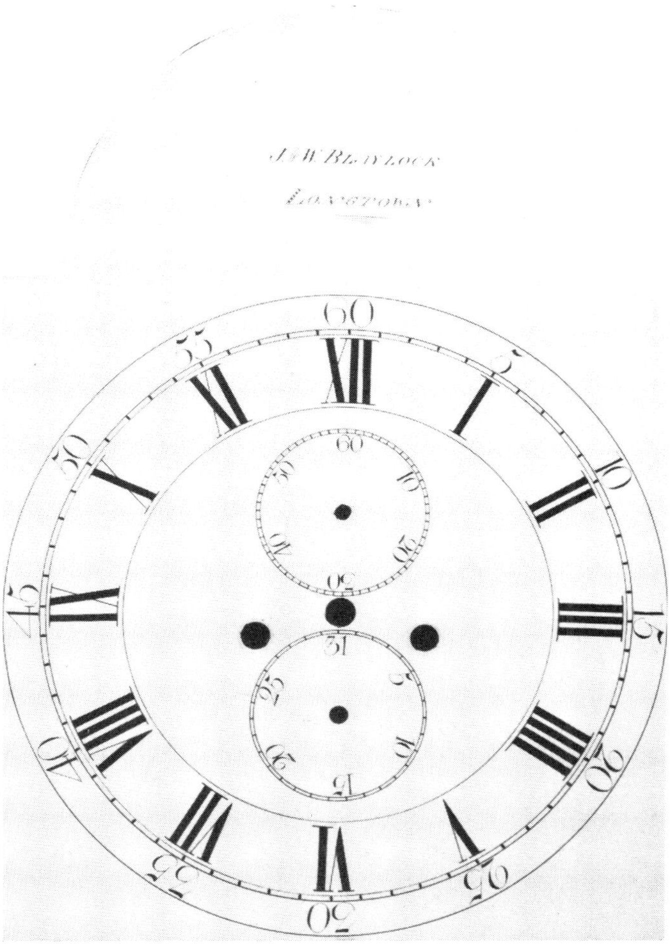

73 An unusually plain early (1791) Whitaker and Shreve–Beilby Hawthorne dial made for William Blaylock of Longtown. See page 124

always done by the same artist. The false plates of these are of all four types. Sometimes the girl is in the break arch, such as this one (**Illustration No. 74**) made for Robert and Ann Dobson *c.* 1805, who the author suspects were owners of the clock and not makers. In other cases the girl appears in the corners. The Dobson dial has a false plate with Beilby Hawthorn top and W.W. Halifax bottom. It is interesting to note that the ladies in the corners of this dial have disproportionately large feet. These large feet have turned up in at least one other dial. The girls in pink were also used after 1805 when the Arabic numerals became more common. Another feature of these Halifax–Newcastle dials is the use of gold leaf edge and chapter ring border framing which in

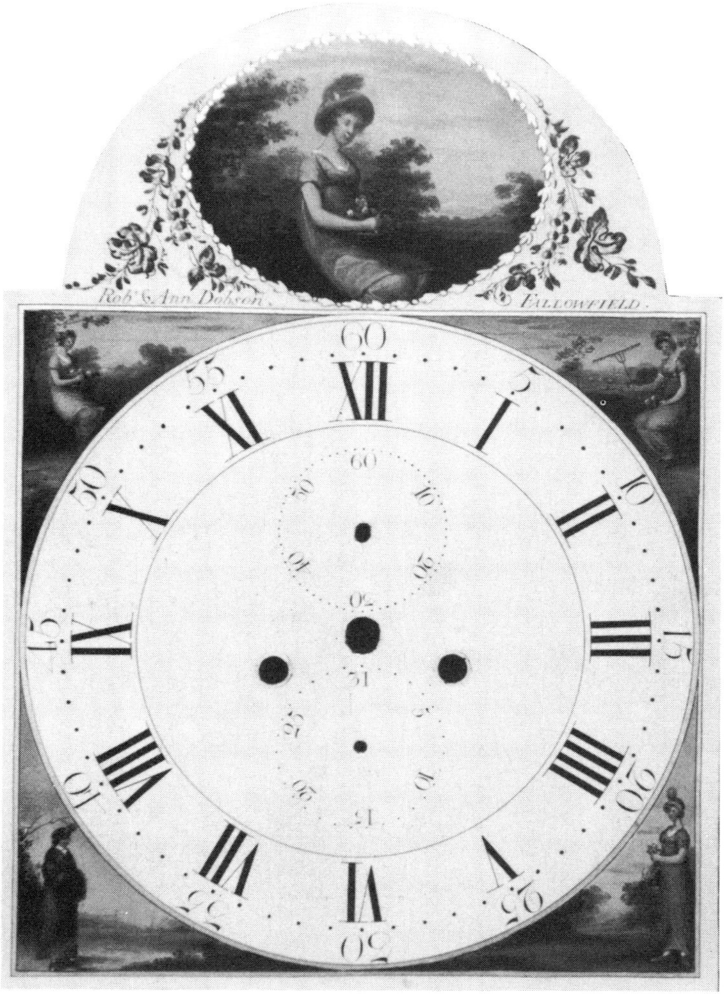

74 Beilby Hawthorne and William Whitaker dial made for Robert and Ann Dobson of Fallowfield *c.* 1805 with the typical 'girl in pink' on the break arch. See page 125

some cases allows the chapter ring circle to protrude beyond the square edges. This can be seen in the completely floral type of dial (**Illustration No. 75**) made for R. Mason of Kelso (*c.* 1800). These floral dials often featured auriculas, and the painting tended to be rather tight and stiff. One of these floral corner dials with gold framing is illustrated on p. 71 of Brian Loomes's *White Dial Clocks*.

A very different variety of corner was the fan type (**Illustration No. 76**) which was made for William Watson of Blackburn

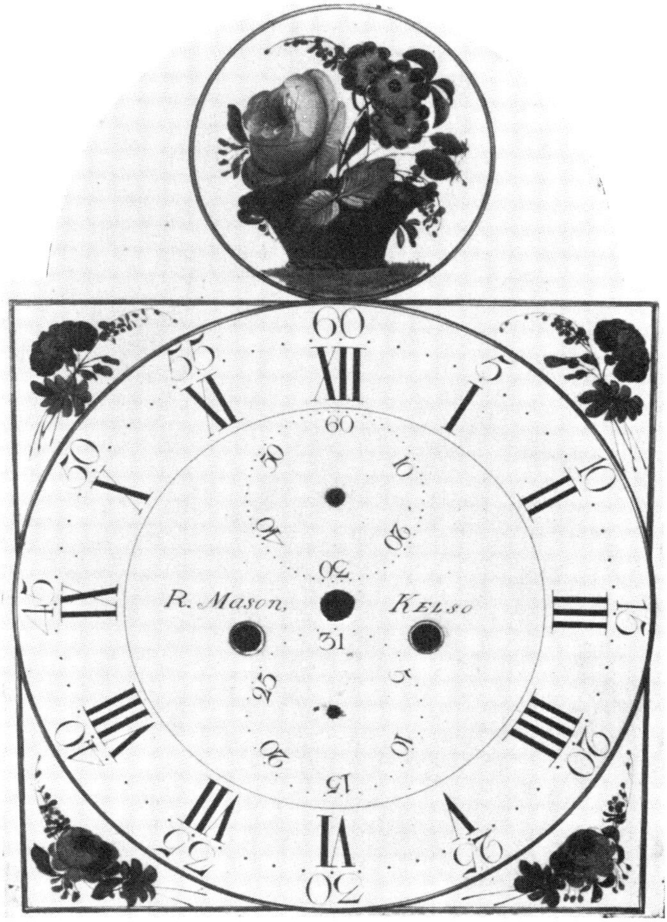

75 Whitaker and Shreve–Beilby Hawthorne all floral dial made for R. Mason of Kelso *c.* 1800. There are auriculas in the break arch flower basket. These flowers often appear on dials from this firm. See page 126

(1790–1814). This dial probably dates from 1800–05, and the fan corners were done in quite a bright shade of green as well as using gold leaf. Some of the moon dials from Halifax and Newcastle had rather pop eyed moon faces, and many featured rather improbable rock formations (**Illustration No. 77**). The cast iron false plate had Wm. Whitaker – Halifax on it, though other fan corner dials had different types of false plate or none at all, using four long feet. The fan corner was also used when the graphics changed to all Arabic numerals. Sometimes, as previously mentioned, a dial can be dated by the event portrayed. This dial (**Illustration No. 78**) with modified fan corners depicts a famous

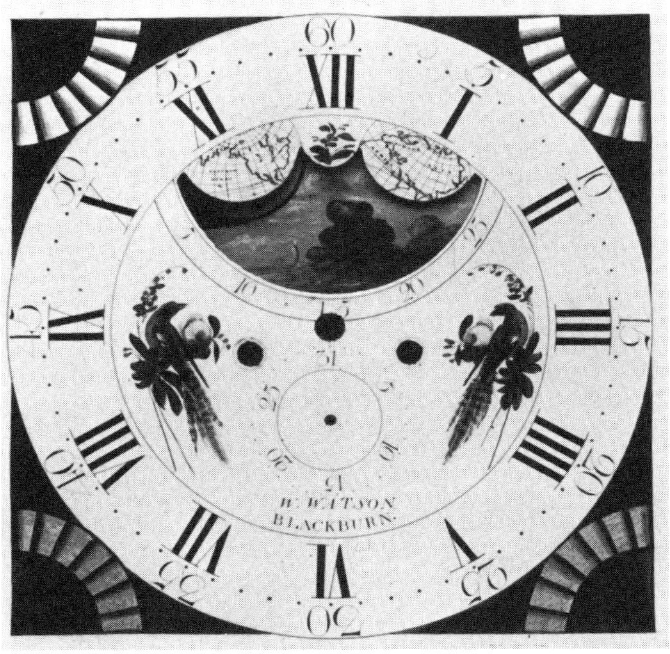

76 William Whitaker of Halifax dial with fan corners in bright green, made for William Watson of Blackburn 1800–5. See page 126

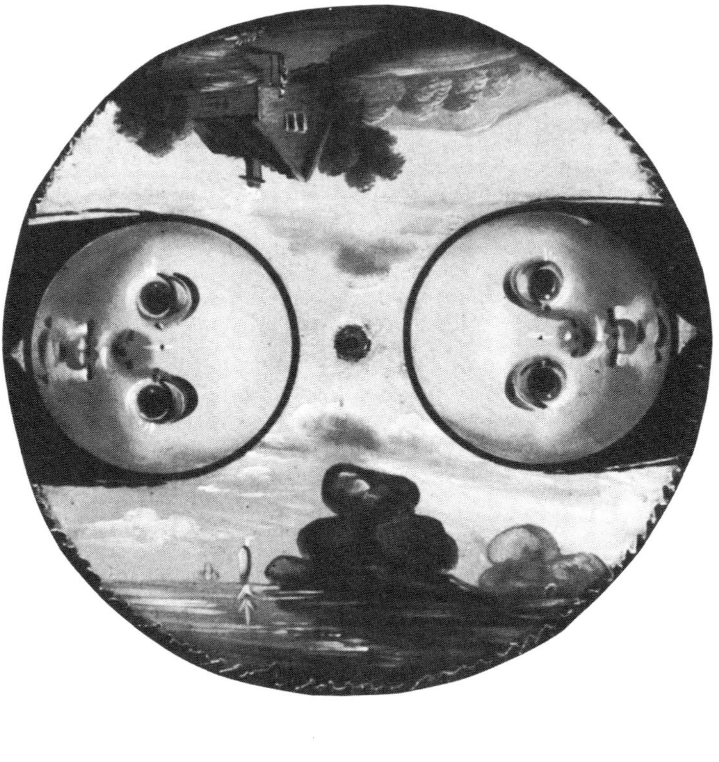

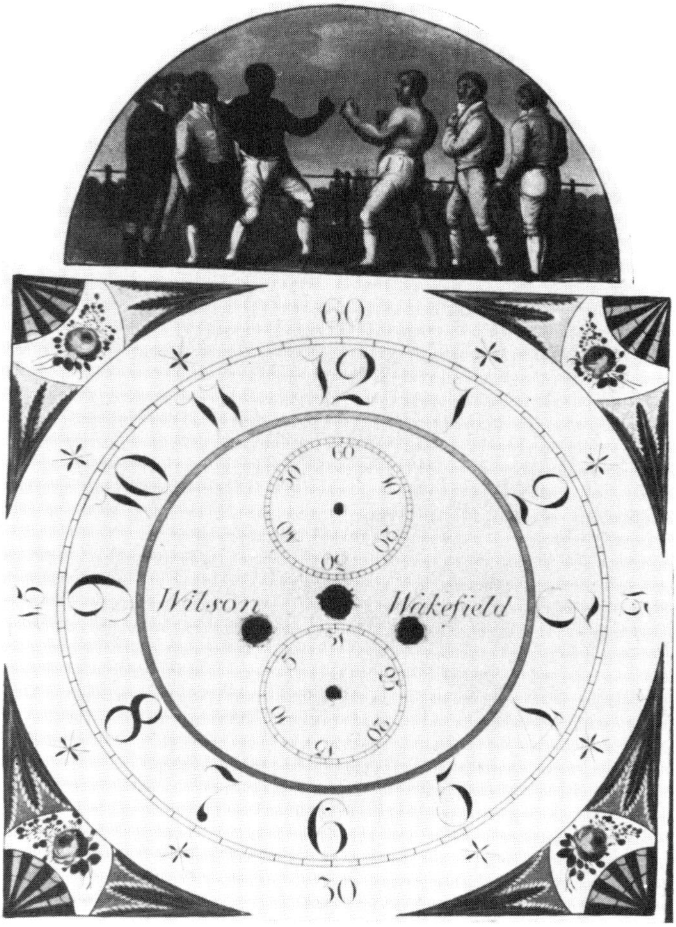

78 A Halifax–Newcastle dial with modified fan corners and a famous prize-fight of 1811 in the break arch. Made for Wilson of Wakefield and was probably a special order. See page 127.

prize fight in 1811 between the English champion John Cribb and John Robichaux from America. There were two such fights that year and Cribb won both of them. This dial had no false plate, but long feet. Another topical dial was made with a break arch scene depicting a couple where the man is wearing a military uniform and there is a small crown on a cushion in front of them. This is probably meant to celebrate a Royal engagement, and it was made for Richard Morland of Kirby Malzeard.

One interesting point is that the graphics part of the Morland dial had to be scraped off, unfortunately, as no undercoat had ever been used and the paint was flaking. Imagine the author's surprise to find the graphics and corner decorations had all been scribed into the metal surface. This is

---

77 This is the pop-eyed moon belonging to the Watson dial. Note the curious rock formations. See page 127

unusual, to say the least, but could give a clue as to how these dials were planned. A tracing would have had to be taken from the scribed marks in order to transfer this to the dial once the base paint was applied.

The floral type of corner changed, probably about 1805–10, to a much more spectacular kind, using a base of silver leaf with translucent coloured varnish paint, often in red or green which gave a glowing effect. Gold leaf was sometimes based in conjunction with the silver leaf, such as in this dial (**Illustration No. 79**) made for John Alker I of Wigan (1797–1832). Though the Alker dial was probably 1805, the same type of corner continued to be used in Halifax and Newcastle up to 1817 when Shreve died and the firm Whitaker and Shreve went out of business. After that the silver leaf corner probably was used by other firms in that area. There were other dial makers listed as working in Newcastle, such as William Foggan (1820–40), James and William Robson of Drury Lane (1838–58), and John Denniston of Halifax *c.* 1834 but, unfortunately, none of these makers appears to have marked their dials. However, it is certain that dials were produced in this area

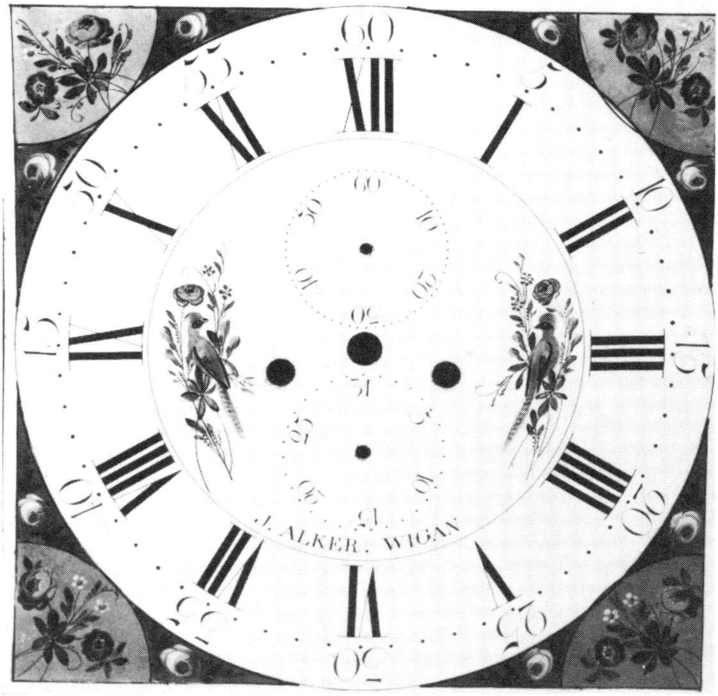

79 Silver and gold leaf to the corners of this Halifax–Newcastle dial made this very eye-catching indeed. The clock maker was John Alker of Wigan, probably John I *c.* 1805. See page 130

after 1817.

Not all the graphics of the Halifax–Newcastle dials were as elegant as the Alker one. The silver leaf floral corner was used with Arabic numerals which could be very heavy indeed. Later there were Roman numerals with very small quarters, and lastly the Roman numerals alone. There were variations on the type of corner, but it nearly always included some of the small pink roses seen in the Alker dial. Nearly all dials of this type using silver leaf have some deterioration of the silver resulting in a great multitude of fine cracks at best, and at the worst a complete loosening of the paint so that it comes off at a touch. Another variation of the silver leaf corner was one incorporating a silver leaf band coloured gold with roses on it nearest to the chapter ring, and the rest of the corner red or green with more roses on it. A good example of this type can be seen in Brian Loomes's *Grandfather Clocks and their Cases*, p. 227.

Twelve o'clock moons were popular in this area, and several of them were 30-hour ones employing false plates and small dummy winding holes. Many of these 12 o'clock moon types of dial had maps on the small hemispheres, and the maps used were different to those used in Birmingham. Garlands of flowers also were popular in this position, sometimes employing silver leaf. More than one type of moon dial painting was used by these Halifax–Newcastle dial makers. Not all of the moons had the pop eyed look, and some of the moon dials had houses with distinctive red roofs. These houses also appeared in the corners and break arches, and were used in both the Arabic type of dial and the all Roman variety. The author has restored one dial of this type, made for Edward Lewthwaite of Halifax (1850–71) which had Roman numerals and long feet. Another earlier dial with Arabic numerals and quarters is illustrated in Brian Loomes's *Grandfather Clocks and their Cases*, p. 244, where the houses in the corners are very similar to the Lewthwaite ones but the scene in the break arch is quite different. That particular clock dial was made for Robert Bartliffe of Malton (1823–55).

The four seasons dials changed over the years from the girls in pink dresses to a quite different style of lady, as seen in this dial with no false plate made for Joshua Bates of Huddersfield (1814–37) (**Illustration No. 80**). The style of dress indicates that it was probably made in the 1820s, allowing for the fact that dial painters may not have kept up with the latest fashions. The painting of both main dial and moon dial is a bit sketchy, and the Georgian mansion on the moon dial looks definitely askew, though possibly this was thought not to matter as the moon dial would revolve anyway. This particular moon dial was iron, and quite a few of even the early Halifax–Newcastle moon dials were iron instead of the more usual brass. The style of the 'signature' on the Bates dial is very distinctive, especially as regards the scrolling. A number of these Halifax–Newcastle dials use this type of scrolling, which is instantly

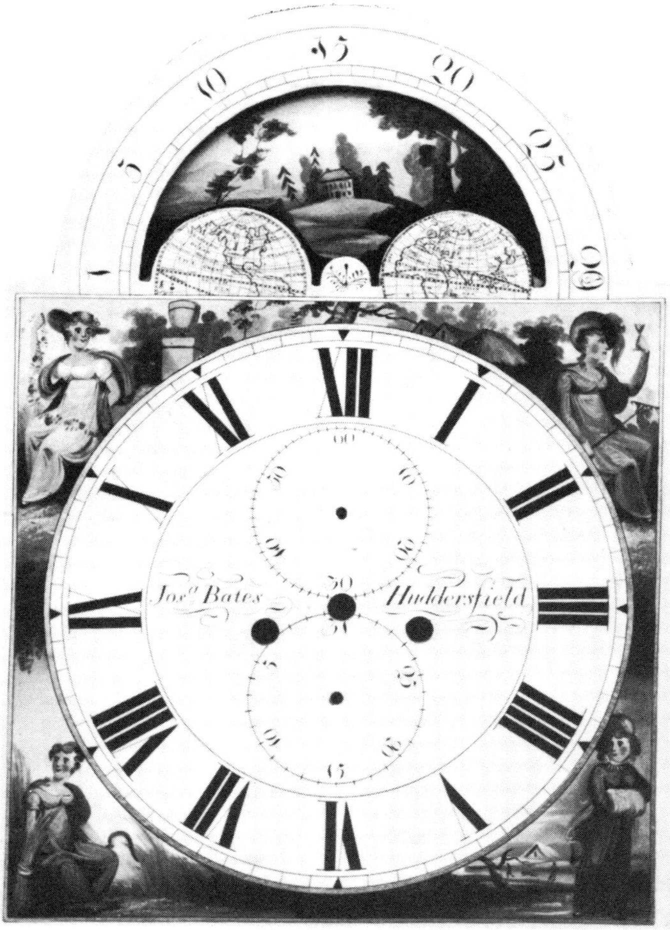

80 A later (1820–5) type of Halifax–Newcastle dial with long feet. The corners represent the four seasons and are rather sketchily done. It was made for Joshua Bates of Huddersfield. See page 131

recognizable as being very different from the Birmingham variety. A complete contrast is provided (**Illustration No. 81**) by this beautifully painted four seasons dial made for William Rowley of Liverpool (1830–48), which has an unmarked sheet iron false plate. The house on the moon dial was painted with a red roof, and the maps are the same as were used by Whitaker and Shreve/Beilby Hawthorn.

Dial makers, however remote, did not exist in isolation from each other, and trends passed from one dial centre to another, though certain areas of the country had favourite styles. The vogue for a coloured chapter ring has been mentioned before (**Illustration No. 59**), but it

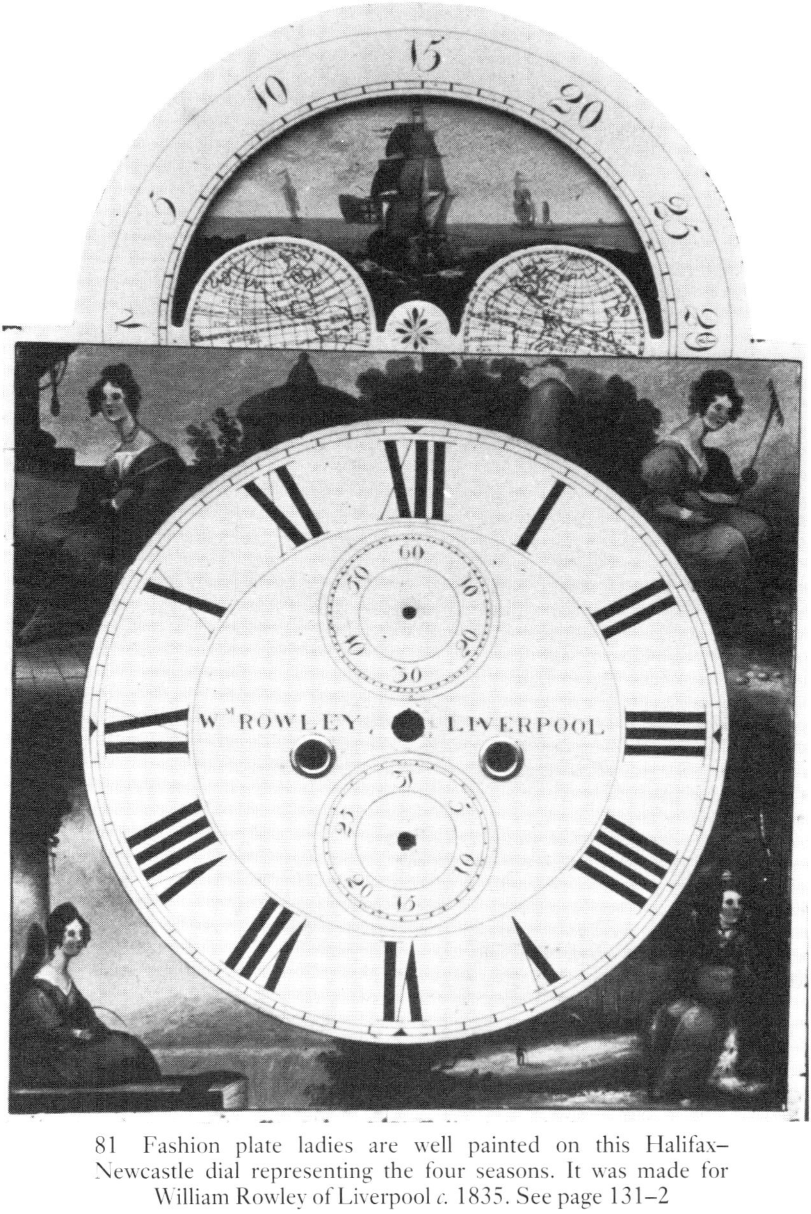

81 Fashion plate ladies are well painted on this Halifax–Newcastle dial representing the four seasons. It was made for William Rowley of Liverpool *c.* 1835. See page 131–2

could have appeared as early as 1825. There is a dial on p. 117 of Brian Loomes's *White Dial Clocks* made for one of the Stonehouse family of Leeds, where the chapter ring is black with white rather heavy Arabic numerals and quarters marked outside the black area. The style of the

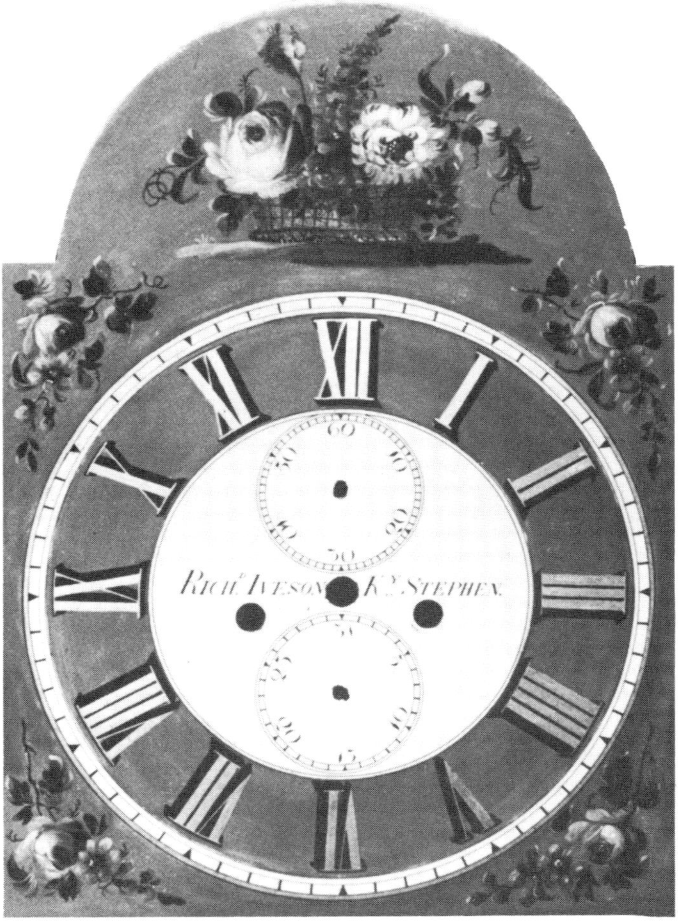

82 Although no maker's mark of any sort was on this striking dial with an all-blue background dial, it is a Halifax–Newcastle dial of about 1840, made for Richard Iveson of Kirkby Stephen. The numerals are in gold leaf and outlined in black. See pages 134–5

'signature' is very similar to the Bates one, and the dates given are 1820–30. The author believes this to be a Halifax–Newcastle dial, though the corner paintings of the Four Virtues are unusual as is the break arch subject 'The Three Nuns'. A development of the coloured chapter ring was the all-coloured background to the whole dial, except for the area in the centre of the chapter ring (**Illustration No. 82**). This dial, with its mid-blue background colour and gold black shadowed Roman numerals, was made for Richard Iveson of Kirkby Stephen (1828–69) and probably dates from about 1840. The same blue colour

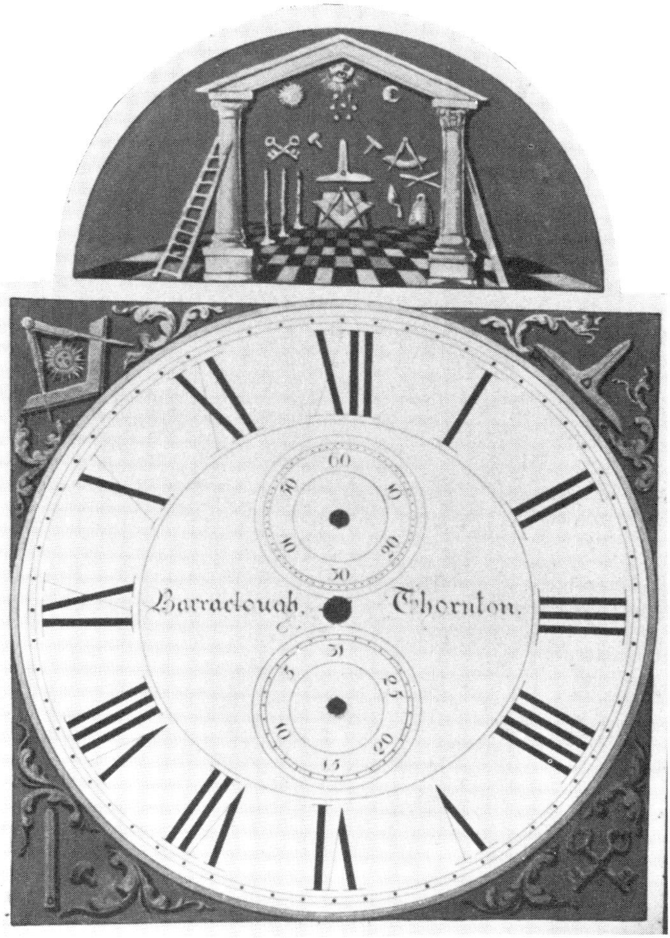

83 The same blue as mentioned in Illustration No. 82 predominates in this 30-hour dial made for a masonic temple. The clock maker was one of the Barracloughs of Thornton *c.* 1845. See page 135

predominated in what was probably a Special Order dial (**Illustration No. 83**) made for John Barraclough II of Thornton (1837–80). It could have been made for a masonic temple in the 1840s.

It can be seen that this area produced quite a number of dials which are usually easily recognizable, spanning the years 1795–1840. More research ought to be done on those dial makers after 1817, but the absence of any dial makers' marks makes this rather difficult.

WILLIAM HELLIWELL OF LEEDS

In any area of manufacturing there are bound to be small independent makers whose products seem to be few and not well known. One such firm was William Helliwell of Leeds (1815–53). Besides producing dials he also made or put together movements, so he did not use any false plates. One of his dials is most unusual in that it depicts manufacturing processes, including a woman working a fly press, which is painted in the break arch. This dial was made for an unrecorded clock maker, J. Roberts of Leeds, though there was a Charles Roberts in Leeds (1797–1832). The dial can be seen on request in Newark Houses Museum in Leicester. One thing noticeable on Helliwell dials is the unusually wide minute band. Another Helliwell dial had 'W. Helliwell, Leeds' on the front of the dial and the back plate of the movement was stamped 'W. Helliwell Leeds 1822'. The movement appeared to be batch produced and did not have any of the scribing lines usually seen in the individually made product. This particular dial was not unusual in any way, having cottages and castles in the corners, and the break arch contained a painting of Sterne's Maria and her little dog from *A Sentimental Journey*.

Brian Loomes records as dial painters in Leeds a Sarah Whittacre (1834–7) as well as James and John Schwanfelder (1810–37), but so far nothing can be definitely attributable to these dial makers or painters.

BARNSLEY HARROGATE DIALS

Sometimes two or more distinctive dials turn up that were obviously made by the same person or firm but, being unmarked, nothing can be found out concerning the dial makers. One of these was made for William Mosley of Pennistone, near Barnsley (1822–33). The dial sheet was thin, with a thin ground paint, and rather crudely painted fan corners, flared Arabic numerals and a fighting cock in the break arch. It was made for a 30-hour clock, and below the centre hole were distinctively painted roses. These same roses appear on another 30-hour dial made for John Fothergill of Knaresborough near Harrogate (1807–51). The break arch contained a memorial to Nelson, and there are small Arabic quarter numerals as well as the main flared Arabic numerals. This dial probably pre-dates the first one by a few years but both these dials appear to be 1810–20. The Fothergill clock can be seen on p. 105 of Brian Loomes's *White Dial Clocks*. Neither one of these dials have stood the test of time very well. In fact the Mosley dial had to be completely stripped and repainted due to rust damage and the lack of a priming coat.

SCOTTISH PAINTED DIALS

Felix Hudson has written a very complete analysis of Scottish Painted Dials in four articles (1985–6) for the *Antiquarian Horology Magazine* but as the reprint of this may not still be available it might be useful to comment on the subject. Also Brian Loomes has a short chapter on Scotland in his book *White Dial Clocks*, as well as a section on Scotland in his *Grandfather Clocks and their Cases*.

Birmingham-made dials were used in Scotland at nearly all times during the period 1770–1870, but the Scottish-made product was usually distinctive. The square dial did not seem to have been used, nor was the 30-hour clock very popular. Date apertures of the semi-circular type did not appear on Scottish dials, and both false plates and long feet were used. The round dial was quite common from 1820–60. Break arch moon dials appeared in imported Birmingham dials, but these were not all that usual. When they were used the moon dial often had a band incorporating tidal numerals on it, and there was usually a pointer of some sort. Scottish-made dials with moon works were uncommon, and mostly confined to the pre-1820 period. Twelve o'clock moons were not used. Some bracket clocks with Birmingham dials existed in the 1785–1800 period, but the longcase clock definitely was the most fashionable in Scotland. Very few Scottish-made dials featured automata.

The earliest Scottish-made dials were very similar to the early Osborne ones with all gilded gesso work, but they were done in a slightly heavier and more elaborate style. The firm of William Dalloway of Edinburgh had the monopoly of dials in Edinburgh from 1780–1800. His son carried on the business after William died some time around 1793, and the firm continued into the 1810–15 period. This firm used cast iron false plates. Another name was T. Smith and Stephenson, Edinburgh *c.* 1815, as well as James Patterson of Perth. Robert Somerville or Somerveil made dials in Glasgow from 1798–1804. There were certainly many other firms who made dials in Scotland, and a complete list of all Scottish dial makers known at the present can be found on p. 57 of Brian Loomes's *White Dial Clocks*. Edinburgh dials tended to have high break arches, while those made in Glasgow or thereabouts were lower up until 1820. The dial size in Edinburgh before 1800 tended to be small, 12″ at first and later 13″.

The Birmingham dial makers seem to have taken full advantage right from the beginning of the expanding Scottish market for painted dials. Wilson, Osborne, Byrne, Owen, Walker and Finnemore, Walker and Hughes as well as Wilkes and Baker all made dials that went to Scotland, though not all their products had the clock maker's name put on at the factory. A good example of this can be found in **Illustration**

No. 32 of an early Osborne dial with a square date aperture and the name 'Robertson, Dundee' slightly amateurishly done.

As mentioned previously, the very earliest Scottish-made dials were of the gilded gesso corner type, but gradually distinctive styles emerged. In general, the graphics style lagged somewhat behind that of the Birmingham dials. The firm of Dalloway and Son of Edinburgh used an unusual type of minute notation on at least two dials. Instead of the minutes going 5–10–15 etc. or 15–30–45–60, they go 10–20–30 etc. This never seems to have been done in Birmingham, though Irish-made dials followed this system. The earliest (1795–1800) dial made by Dalloway and Son was one with their own name on the dial with what appear to be gold leaf grape leaves with grapes in the corners, husk and acanthus in the break arch together with a Latin quotation. The interim markings are diamonds surmounted by an asterisk.

The second dial was made for William Drysdale of Falkland *c.* 1810, and it followed the same notation. This dial looks in other ways very similar to Birmingham work, with floral corners, and a bird together with flowers in the break arch.

Up to about 1815–20 the Scottish-made dial tended to be influenced by the Birmingham style, with floral or geometric corners, sometimes with single figures or a basket of flowers in the break arch, but done in the Scottish style. After this the brilliantly coloured typical Scottish dial came into its own. False plates were gradually phased out after 1820. The use of silver leaf as a ground for decorative translucent overpainting, often red or green in the earliest dials, meant that the dial appeared to glow, even under meagre lighting conditions. This type of painting had been used previously in the Newcastle–Halifax dials but in Scotland the so-called luminous painting became almost universal.

An interesting example of this technique, presumably in its early stages, is this amusing dial (**Illustration No. 84**) made for John Barron of Aberdeen (1801–52). The corners are done in bright red and green, and use a type of fan design, with fruit. The subject in the break arch is obviously a blacksmith, but tucked up in the blazing forge is the Devil himself warming his hands. This dial is probably *c.* 1815. Although by this period Birmingham dials had usually gone over to the all Arabic numerals, these Arabics were not popular in Scotland even on imported dials from Birmingham. An earlier type of dial was used by John Barron *c.* 1801, where there is the full Arabic minute notation of a larger size and floral corners reminiscent of Birmingham work. The continuing use of the quarter markings went on in Scotland far longer than in Birmingham, and the wide minute band containing these was employed well into the 1830s. The double V interim markings were also typically Scottish, as seen (**Illustration No. 85**) on this quite sophisticated dial exported to Ireland. The style of the 'signature' is most distinctive and

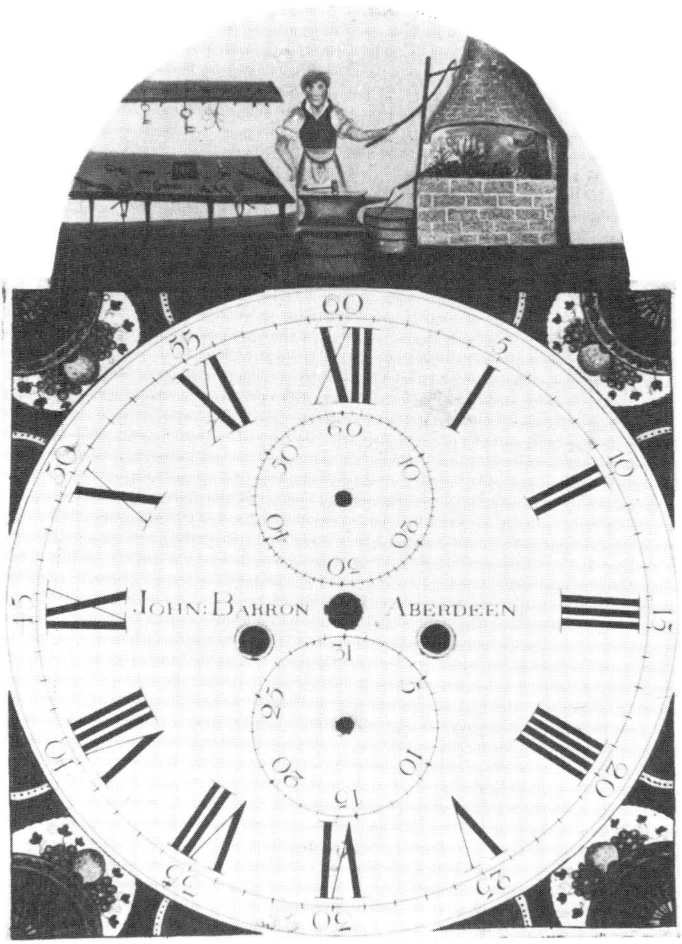

84 An unusual and amusing Scottish-made dial with silver leaf as a background to the red and green corners and a blacksmith in the break arch together with the Devil warming his hands at the hearth. The clock maker was John Barron of Aberdeen *c.* 1815. See page 138

identical to one illustrated on p. 324 of Felix Hudson's *Scottish Longcase Clocks, 1780–1870*. This Dawson Bell of Omagh dial is very colourful, with a liberal use of gold leaf. It probably dates from around 1830.

By far the most popular subject for the corners of these Scottish dials was the four seasons, all appropriately labelled as seen in this dial (**Illustration No. 86**) made for John Gibson of Saltcoats (1825–30). The corners of these four seasons dials, which were quite different from the English ones, tended to have certain features repeated, although painted in varying styles. Thus Spring usually has an urn with flowers

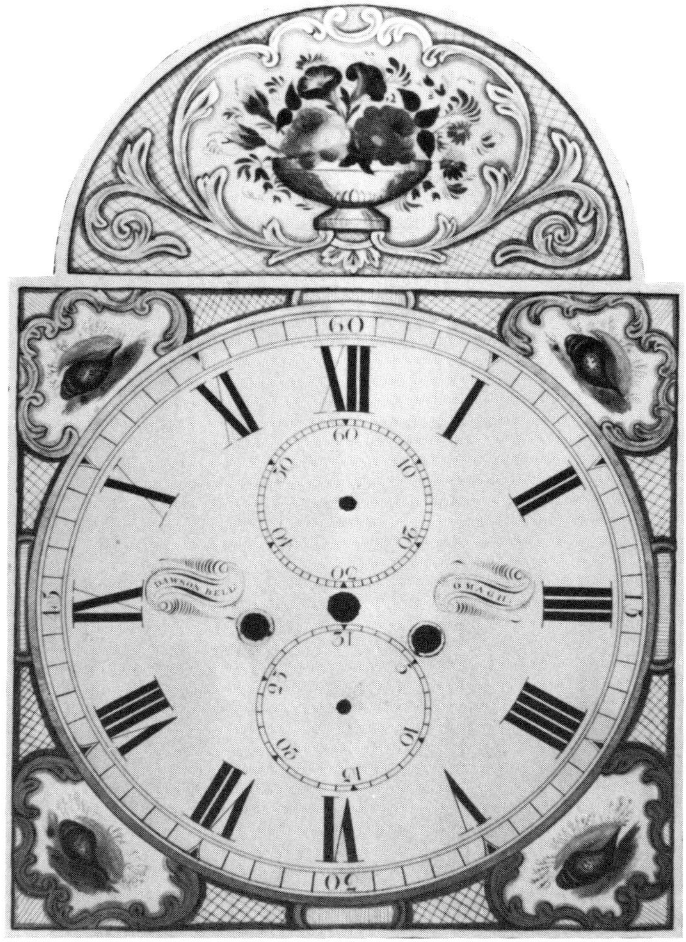

85 Scottish-made dial exported to Ireland *c.* 1830 with very distinctive 'signature'. Dawson Bell of Omagh. See pages 138–9

and Summer a haystack. However, Autumn here is labelled Harvest. The break arch in this period tended to be increasingly important, and dominated the whole dial. 'Farmers visit a Friend' would have been classified as a 'Couthy Subject', but all sorts of other themes were used in the break arches right up to the end of the period in 1870. This Gibson dial has a lot of 'luminous' paint. Although the outlining and details done in black are here relatively light, it could be very heavy indeed on some of the other and possibly later dials.

The Scottish School of Art was very strong from about 1820 onwards, with Edinburgh and Glasgow the main centres. The break arch paintings of many dials were direct copies of well known works, such as John Faed's *The Cottars' Saturday Night* (**Illustration No. 87**), which

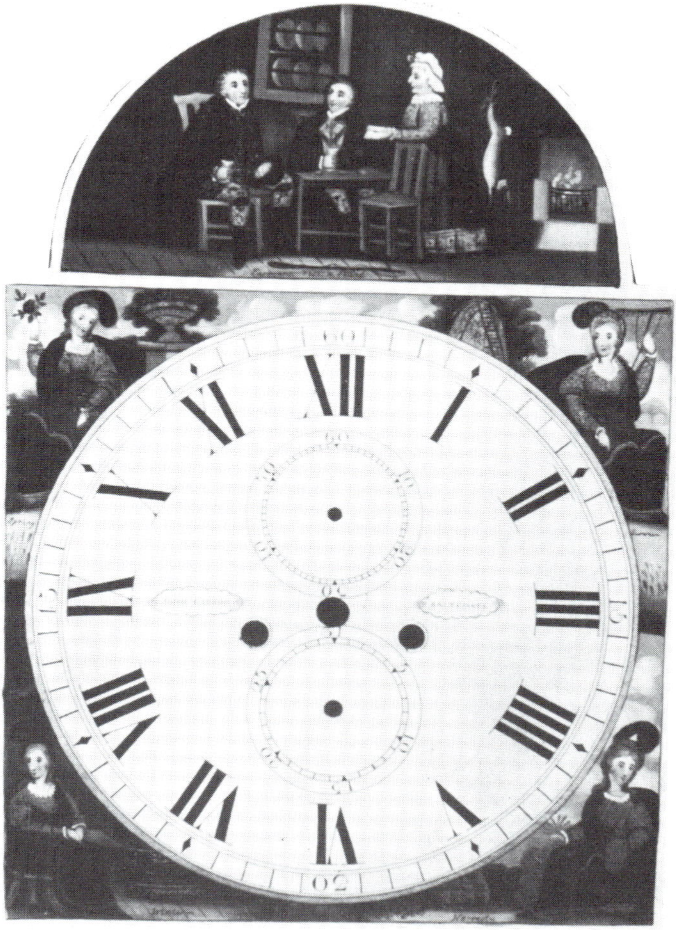

86 Typical Scottish four seasons corners and 'couthy subject' in the break arch. Silver leaf was used under transluscent paint to give so-called luminous efect. This dial was made for John Gibson of Saltcoats, 1825–30. See pages 139–40

appears on a dial made for Daniel Hunter of Lockwinnoch (*c.* 1855). The corner decorations are a Scottish version of the four continents, and again all appropriately labelled. America's bale of cotton is marked 60.

Another distinctive type of corner depicted famous persons, such as this dial (**Illustration No. 88**) made for John Todd of Troongate, Glasgow (1823–37) where Burns, Ramsey, Ferguson and Scott are featured in the corners. Other dials include Douglas, Bruce, Graham and Randolph. This dial also has the double V interim markings seen on the Dawson Bell dial. The break arch scene is 'The Fair Maid of Perth', which is well painted.

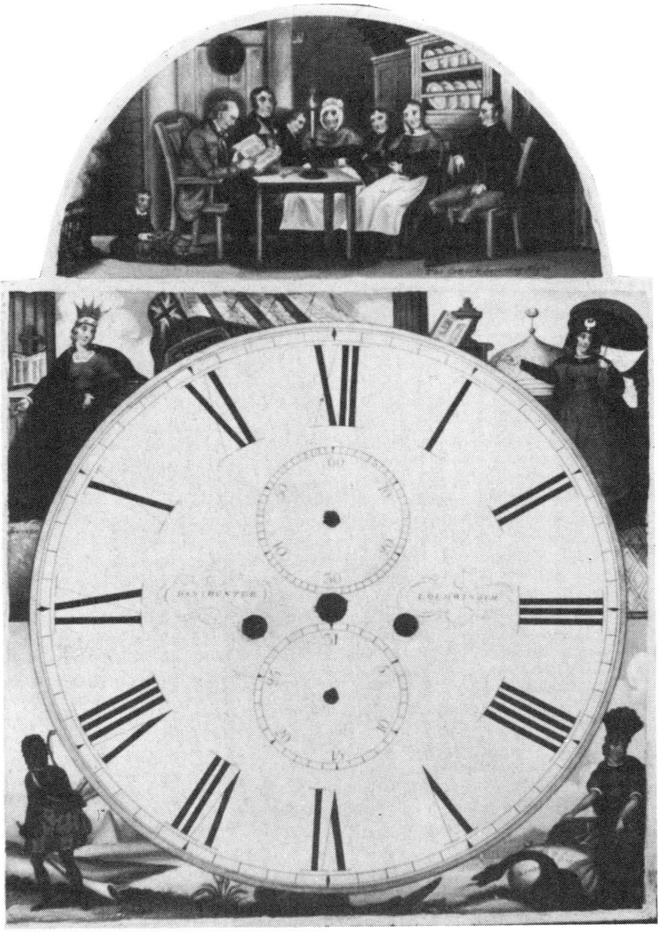

87 Here are the four continents in the Scottish version using 'luminous' colours and 'The Cottars' Saturday Night' in the break arch copied from Faed's well-known painting of the same name. This dial was made for Daniel Hunter of Lochwinoch *c.* 1855. See pages 139–40

One interesting development related to the Edinburgh dials was the use of the slightly curved base to the break arch subject from 1837–70. This dial (**Illustration No. 89**) was made by Peter Bell of Edinburgh for David Peters of 84 High Street, Arbroath *c.* 1837. The back of the dial was marked with the dial maker's name and 'Finest Fancy' done in what was once white paint, which presumably related to the type and cost of the dial. Unfortunately, not all dial makers so marked their work. The corners of the dial are the four seasons, where the ladies are quite fashionably dressed, and the break arch subject is another one of Faed's paintings 'The Soldiers' Return'. Different versions of these paintings turn up in

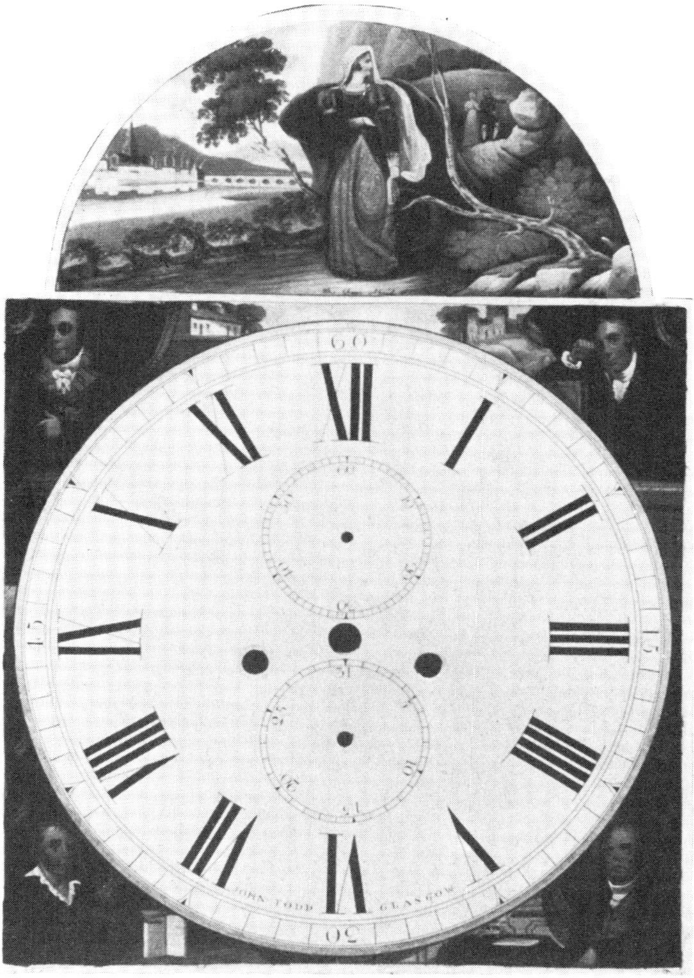

88 Famous personages with 'luminous' paint decorate the corners of this Glasgow-made dial *c.* 1825. The subject in the break arch is 'The Fair Maid of Perth'. The clock maker was John Todd of Troongate. The double *v* markings were used only in Scotland. See page 141

prints of the period, and one of the same subject omits the figure of the mother. Another Edinburgh dial of this curved base type of break arch was made considerably later, *c.* 1870, and featured 'The Lady of the Lake' with a slightly different version of the four seasons in the corners.

Not all Scottish painted dials were well made and the divergence of styles was so immense that it is almost impossible to catalogue all the varieties that still exist. The impression given is that there were some very professional dial makers, probably based in Edinburgh and

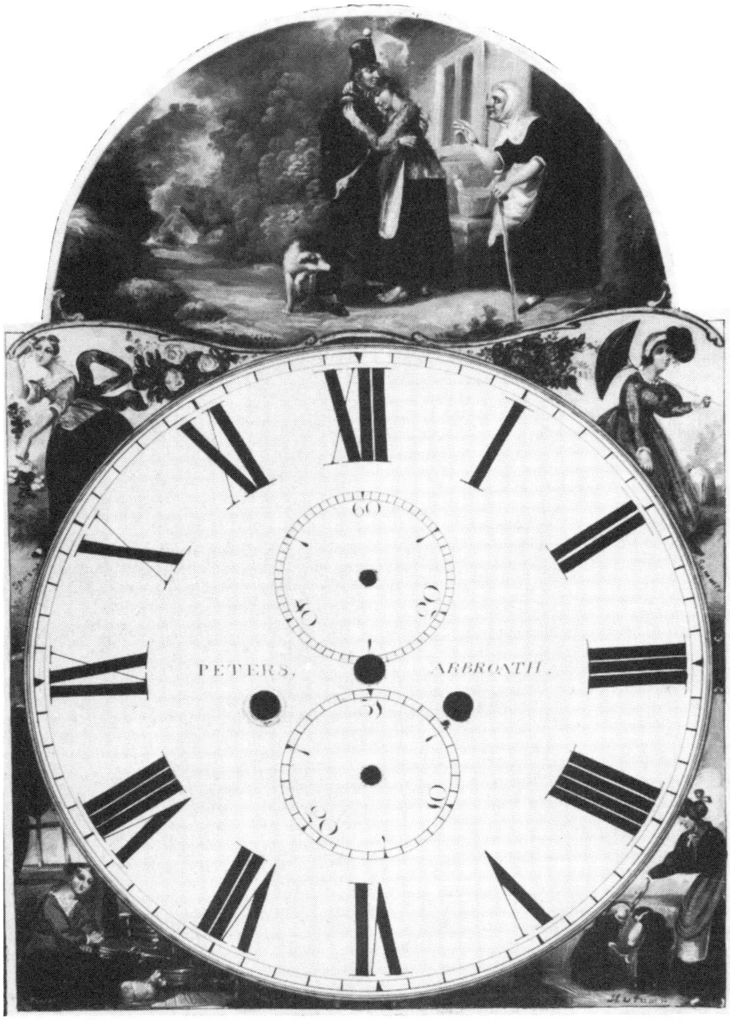

89 'Finest Fancy' Edinburgh dial made by Peter Bell for David Peters of Arbroath *c.* 1837. This is one of the few Scottish dials with the maker's name and type painted on the back. There is a curved base to the break arch illustration, 'The Soldier's Return', which is a copy of Faed's painting. See page 142

Glasgow, but other makers appear almost amateur. This dial (**Illustration No. 90**) is one of the latter type. It was made *c.* 1864 for Robert Aitken of Island Street, Galashiels, and is one of those dials celebrating a special event. During the American Civil War, the export of American cotton upon which the Lancashire cotton mills depended so heavily dried up completely, and starvation was the result. In the north of America, where there had been an abundant harvest, the

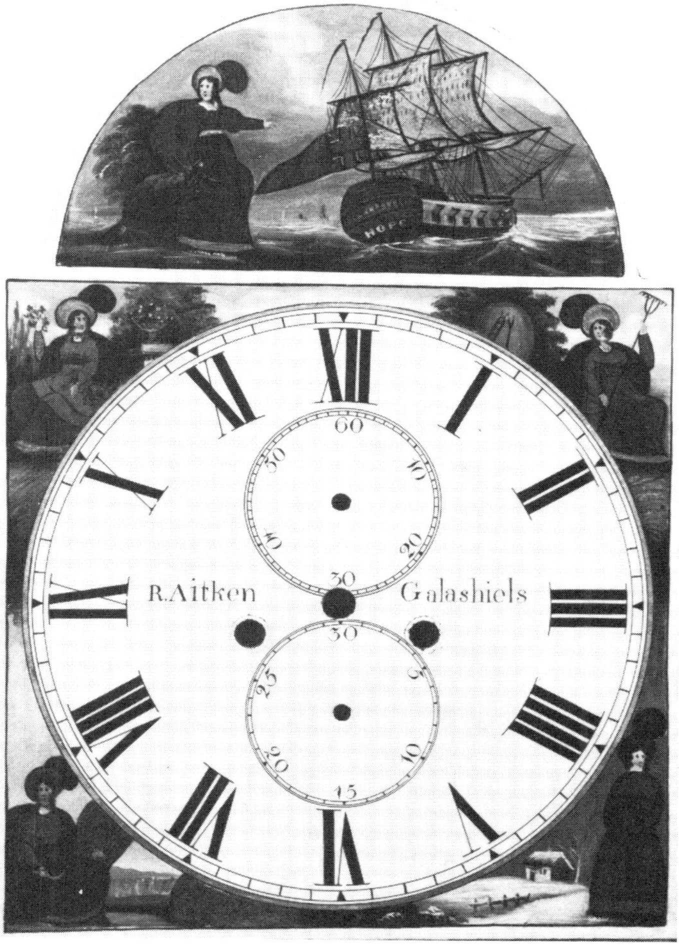

90 A rather crude Scottish dial made for Robert Aitken of Galashiels *c.* 1864 celebrating the arrival of the ship *Hope* laden with flour to ease the famine amongst the cotton workers. See text pages 143–4

steamship Hope was sent to Liverpool, arriving in January 1864 with thousands of barrels of flour to help ease the situation. The author believes it was this event, suitably romanticized, which formed the subject of the Aitken break arch, although the ship painted was certainly from an earlier age. The painting appears quite crude, with heavy black outlining, and the base paint was very thin. These rather poorly made dials have not stood the test of time at all well, and can be seen to be almost visibly deteriorating.

There was one other very unique type of dial only found in Scotland,

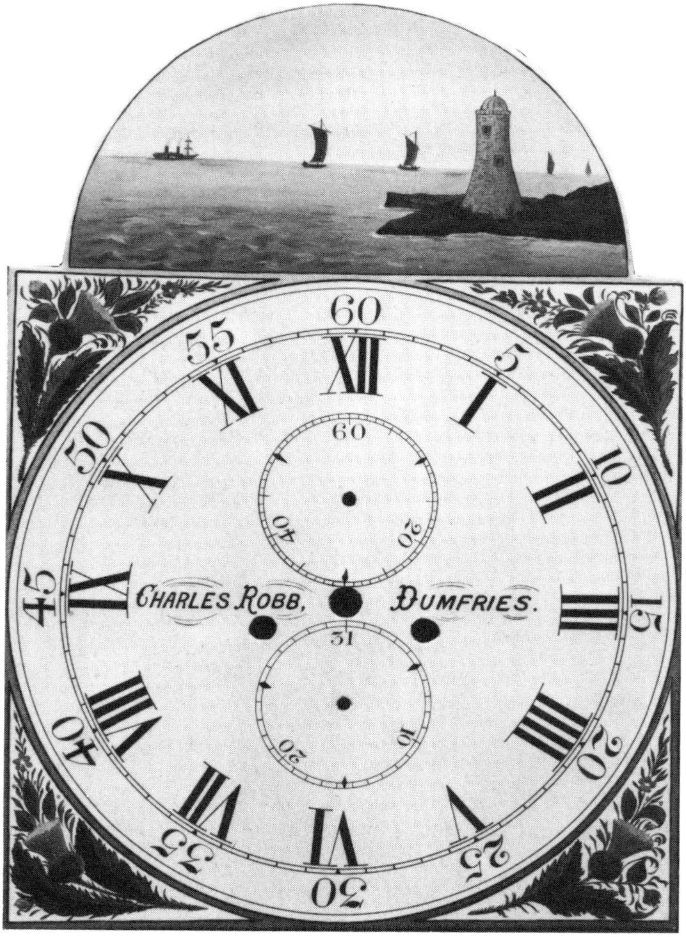

91 This Scottish dial was completely overpainted in about 1845–55. No 'luminous' paint was used. The fully numbered minutes are unusual for this period. See page 146

and that was the round plain dial with either a solid gold chapter ring with black Roman numerals or the later all gold type which had centre decoration of gold worked flowers which almost mimicked the one piece brass dial. This type of dial with gold dated from 1835–1870.

The repainted Scottish dial occasionally turns up where the original was either out of style or badly deteriorated. This dial (**Illustration No. 91**) was probably painted over *c.* 1845–55, though the name Charles Robb of Dumfries is unrecorded. The break arch shows a marine scene rather crudely done, but the fully numbered minute band seems rather unusual at this period though it was used on some Birmingham dials at that time. The type of lettering used for the name is uncommon on longcase dials as well, though it was used on English drop dials. Traces of the original rather crazed base paint could be seen in the corners, but the covering paint was too thick to ascertain what decoration might have lain underneath.

Another unusual thing about some of the Scottish-made clocks was the use of the decorated iron pendulum bob at the time when the silver leaf was used a lot. These decorated bobs used silver leaf and could be floral or geometric, sometimes with a face at the centre, though the decorations did not match the dial. Scottish made dials certainly covered many styles and are quite distinctive, though many of the silver leafed ones have not survived very well. Those that remain in reasonably good condition or are expertly restored give a brilliant effect which explains their popularity in a country where the houses often had small windows and the climate could be gloomy.

IRISH DIALS

It is not generally known that Ireland produced dials from about 1790 or even earlier to *c.* 1820. (See 'The Painted Dial in Ireland' by M.F. Tennant, *Antiquarian Horology*, Summer 1991.) This can be explained in several ways. Even in Ireland these dials were not all that popular, and there appear to have been a limited number made. In many cases they have been taken off the original movement, discarded and spurious brass ones substituted in order to make a clock more saleable. Also these Irish-made dials tended to be unmarked without any dial maker's name so it is difficult to know who really did them. False plates were very rarely used. Again, similarly to Scotland, the Birmingham dial makers exported quite a lot of dials to Ireland in all periods. Some Irish clock makers used both the native and the imported dial during the same period of time. One Scottish-made dial (**Illustration No. 85**) previously mentioned was exported to Ireland in 1830–5, though this could have been unusual.

Dial production appeared to have centred on Dublin but doubtless they could have been made elsewhere as well. Irish-made dials were usually square except those for bracket clocks, which had a break arch. There were four rather stout long feet, and date apertures were not used, though they do appear in the imported dials. As a general rule the early 1790–1810 dials were of very high quality. One of these (**Illustration No. 92**) was made for Thomas Pilkington of Dublin (1795–1824). The standard of the gilded gesso work is very high and the decorated winding holes suggest a slightly earlier date than 1795. The interim seconds markings are very unusual. Another type of early *c.* 1795 dial is this one (**Illustration No. 93**) made for Middleton of

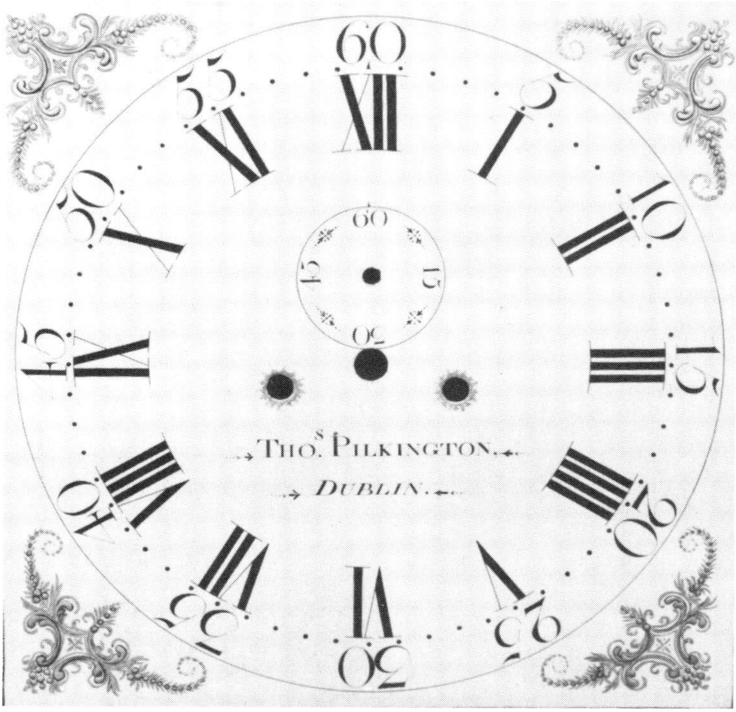
THOs. PILKINGTON.
DUBLIN.

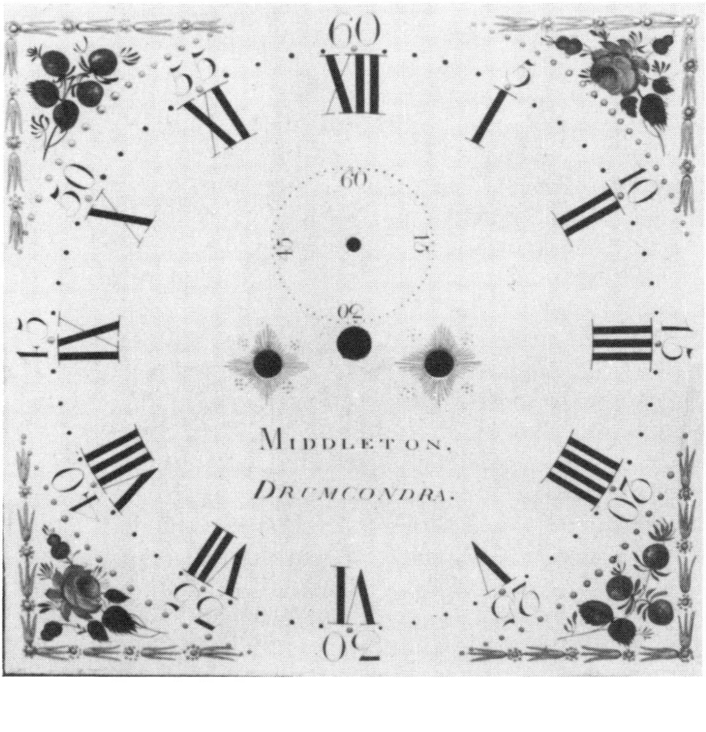
MIDDLETON,
DRUMCONDRA.

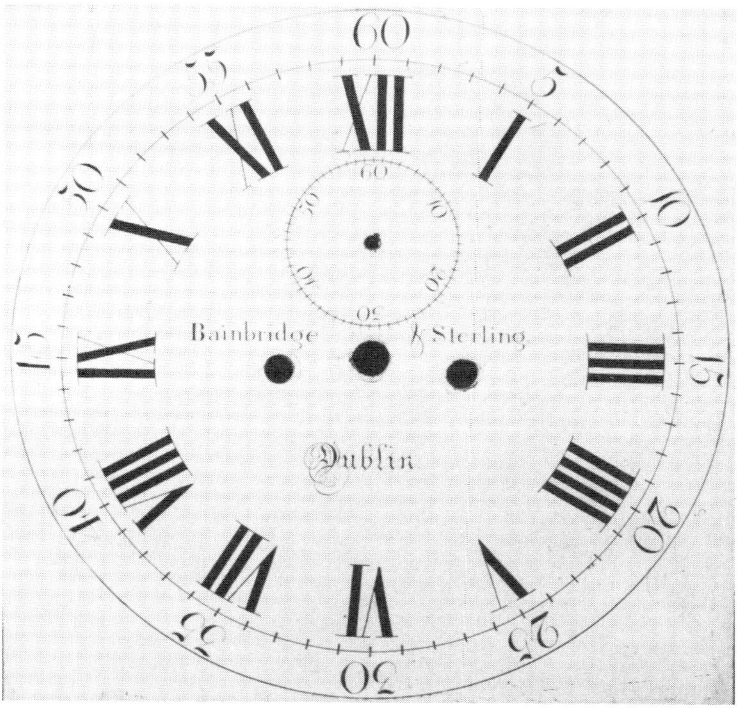

94 A plain style of dial *c.* 1805 made for Bainbridge and Sterling of Dublin with 'Dublin' painted on the back. See page 149

Drumcondra, in the northern part of Co. Meath near to Dundalk. This clock maker is unrecorded but the dial, with its exuberant gilded gesso work and brilliant colours, is eyecatching indeed. The elaborate treatment of the winding holes is unusual.

The colourful style of dial changed to a plainer one about 1805, as seen in this example (**Illustration No. 94**) made for Bainbridge and Sterling of Dublin (1806–9). The minute markings have changed from dots to a barred circle and the Arabic numerals show flares on these ones only. The reverse of this dial had 'Dublin' on it, done in what looked like ground paint. Large brass grommets were used. The effect of this dial is very elegant and restrained. At the same period Bainbridge and Sterling imported a very fine Osborne dial (**Illustration No. 95**) which had the 'New' all Arabic numerals on it and only quarter minute numerals. The name and city were put on in Dublin, and the 'D' in Dublin is very like the Irish-made dial. The gilded gesso corners and break arch decorations are especially fine.

---

92 Irish-made dial *c.* 1790 with elaborate gilded gesso work and gilded winding holes. The clock maker was Thomas Pilkington of Dublin. See page 147

93 Heavy gilded gesso frames the corners of this Irish dial with very elaborate gold work around the winding holes. Made for Middleton of Drumcondra *c.* 1795. See pages 147 and 149

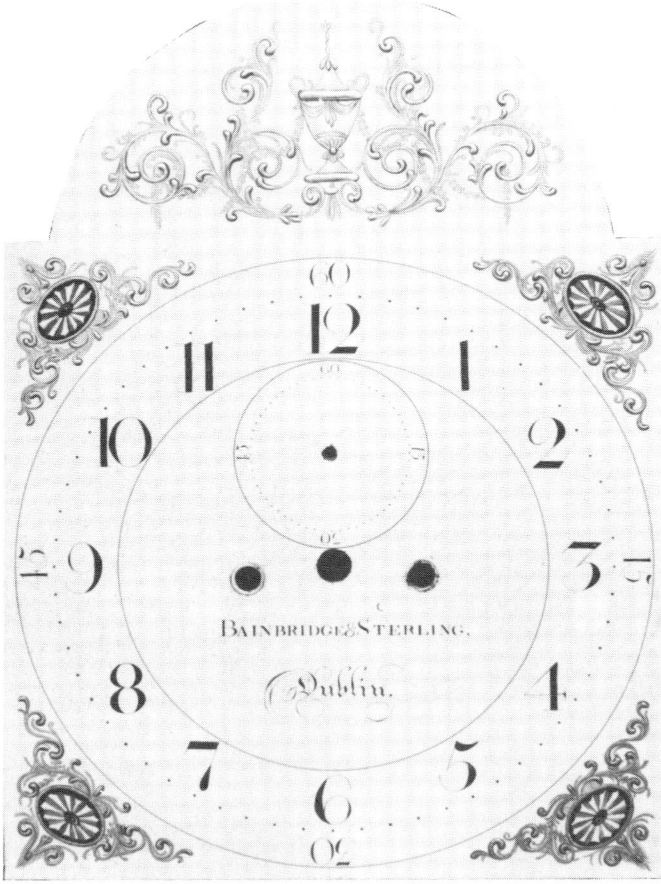

95 An imported Osborne of Birmingham dial made for the same firm as Illustration No. 94 at about the same time (1805). The clock maker's name and town were added locally. See page 149

James Henderson of Dublin (1795) used a plain dial (**Illustration No. 96**) in about 1810 which had 10 minute numerals on it, and gilded gesso dots near the hour numerals. The winding holes were also gilded. A slightly later *c.* 1815 Henderson dial (**Illustration No. 97**) only had minute numerals on the quarters and the corner decoration was done with gold leaf plus a few gesso dots. Crosthwaite and Hodges of Dublin (**Illustration No. 98**) are listed as 1803–11, but this very plain dial with its heavy Roman numerals looks a bit later than 1811. The Edward Smith of Dublin (**Illustration No. 99**) *c.* 1820 dial had very slender

---

96 A rather plain Irish dial *c.* 1810 with minute numerals every 10 minutes and gilded gesso dots. It was made for James Henderson of Dublin. See page 150

97 Made for Henderson with typical Irish corner motifs *c.* 1815. See page 150

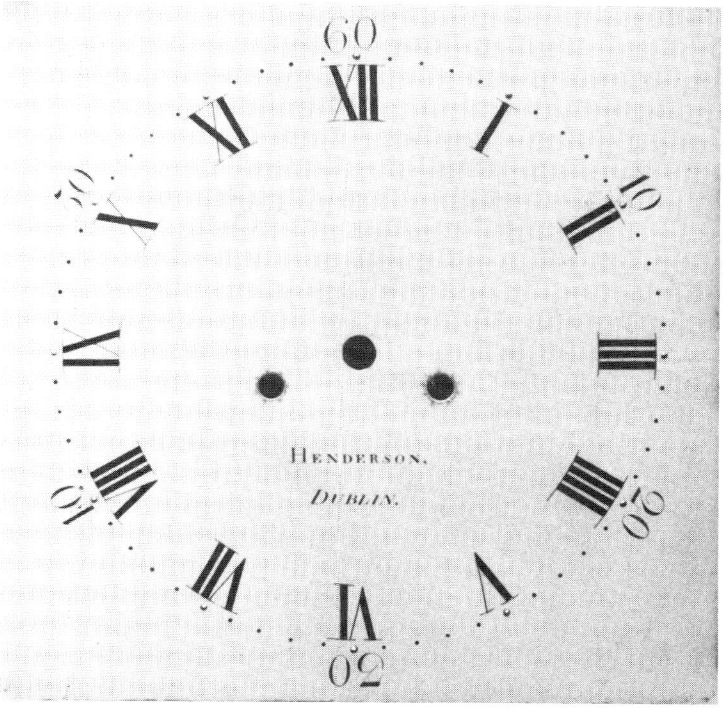

60
Henderson,
Dublin.

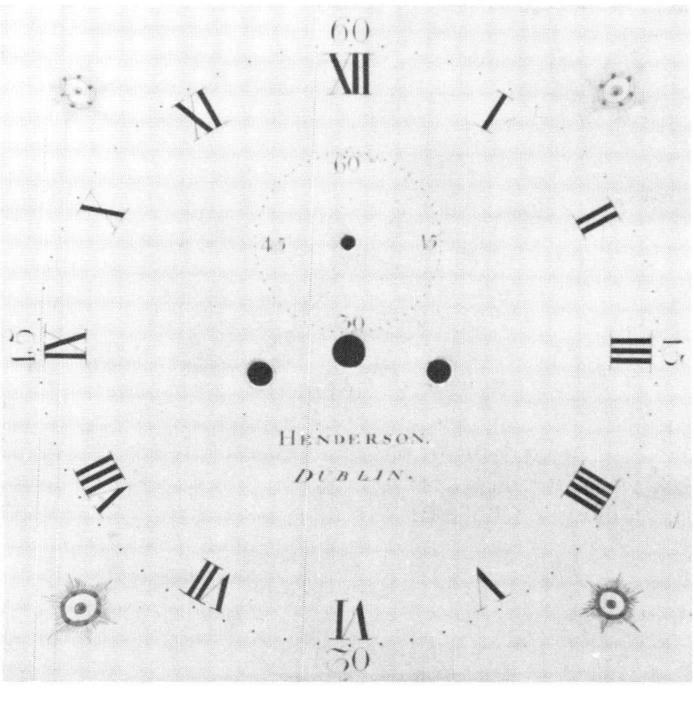

60
45
15
30
Henderson,
Dublin.

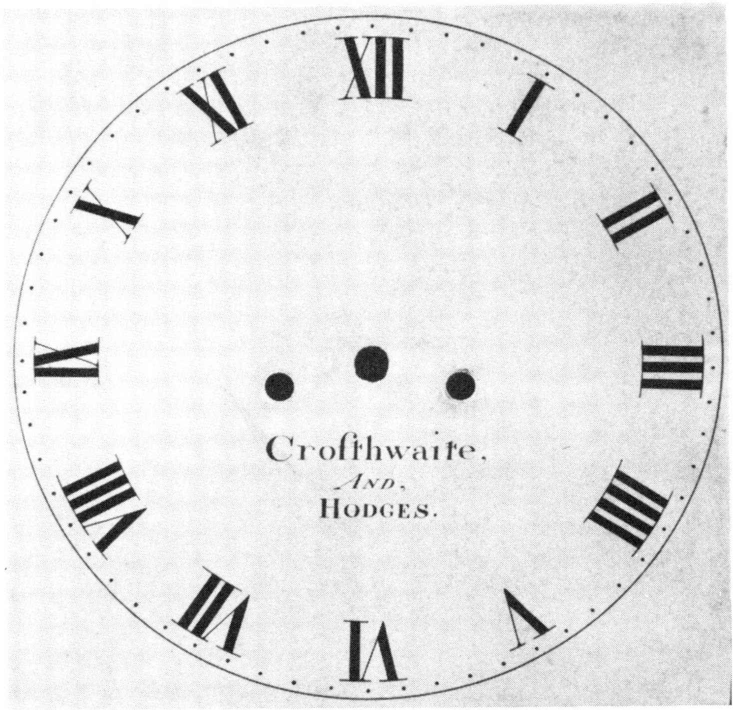

98 A very plain dial probably made *c.* 1815 for Crosthwaite and Hodges of Dublin. See page 150

Roman numerals and corner decorations which are similar to the second Henderson dial. The ground paint was cream coloured and very thin.

One peculiarity of moon dials, both native and imported, in Ireland was that the dials were made of brass which was much thicker than usual, and the teeth were filed. An Osborne moon dial made for Alex Wiley of Lisburn *c.* 1790 or earlier was probably a special order, as no other Osborne has been seen with brass of this gauge. The Birmingham makers who exported to Ireland were: Wilson, Osborne, Francis, Owen and Wilkes, but there were probably others. The moon dials made in Ireland were all for brass dials.

Bracket clocks were also popular in Ireland, both those with Irish-made dials as well as imported English ones. The presumably Dublin-made dials tended to be in the earlier period of 1790–1810. One particularly fine dial with Roman numerals and Arabic quarters was made for Samuel Bailie of Dundalk *c.* 1805. It had a break arch and a lot of very good gilded gesso work. Another slightly later dial, also with a break arch, was made for Crosthwaite of Dublin. It had Roman numerals only, with less elaborate gesso work. The English imported bracket clock dials tended to be very plain, and were often round. Some

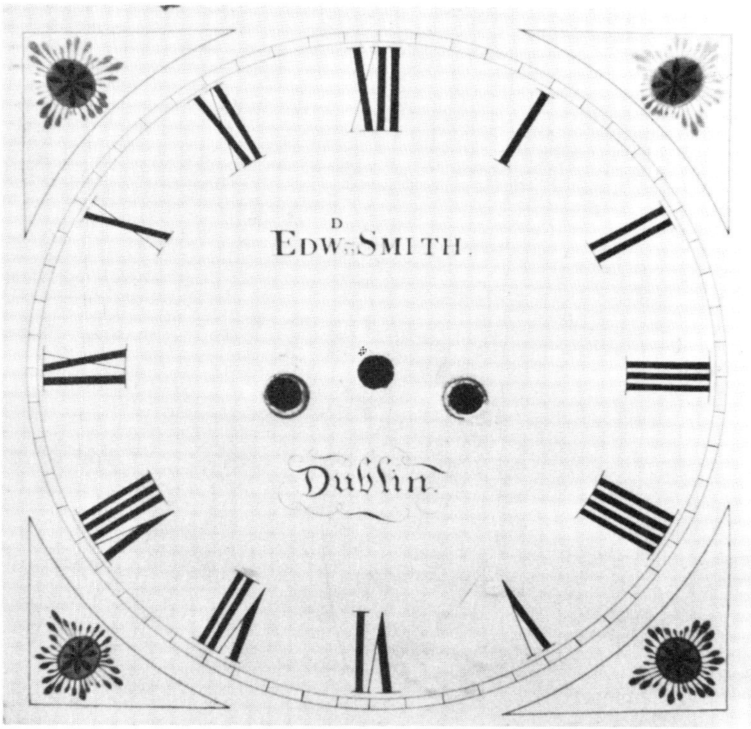

99 This is the latest in the Irish-made series *c.* 1820. It has a very thin dial plate, thin cream-coloured base paint and gold and black corner decorations. It was made for Edward Smith of Dublin. See pages 150 and 152

of them were of the dished variety as well. It seems a pity that Irish-made painted dials should go so unrecognized, especially as their high quality in the early period (1790–1810) matched anything Birmingham could produce.

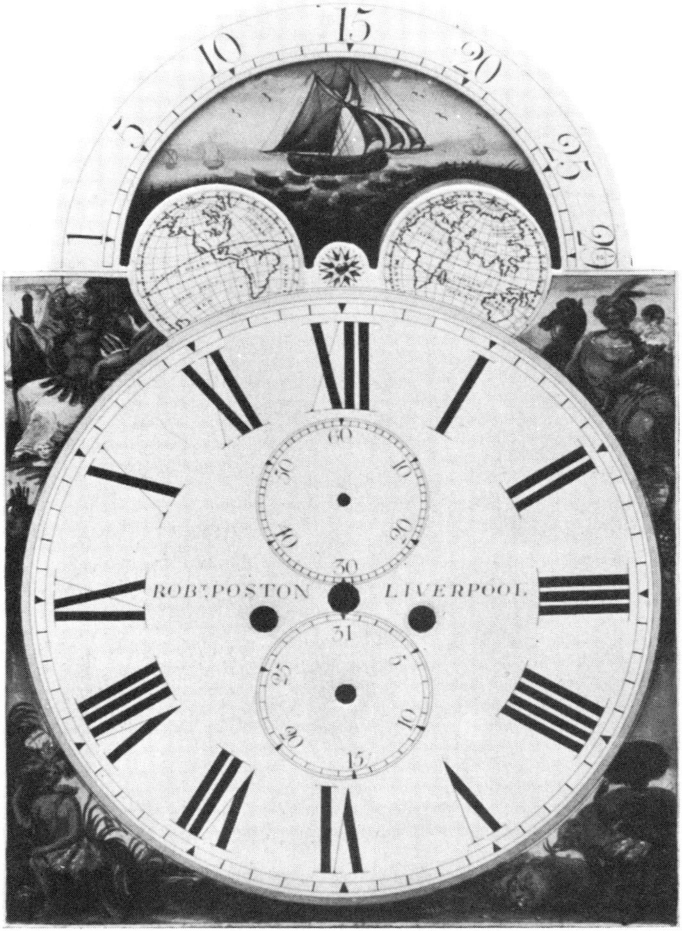

100 Robert Poston together with Frederick Woldnough of Liverpool produced this four continents dial (1824–34). The maps are of very late variety with outside lattitude markings. The false plate was unmarked sheet iron. See pages 154 and 155

LIVERPOOL DIALS

Liverpool was certainly a centre for dial making, though many of its products have gone unrecognized because they were not marked in any way. Only two makers are recorded who did mark their dials, and both appear to have worked in the later period after 1820 when even many of the Birmingham dial makers did not bother to mark their wares. Probably the earliest dial makers in Liverpool were the Poston-Woldnoughs. Robert Poston's dates are given as 1824–59, and he was listed as a clock and watch dial maker. In *Gores Directory* for 1827 the firm Poston and Woldnough 1825–41 was listed as watch dial makers, but by 1829 Robert Poston was working alone as a watch dial maker, while Frederick Woldnough, at a different address, was listed as a watch and clock dial enameller. The probability is that Woldnough made the clock dials, though this large and colourful clock dial (**Illustration No. 100**) has Robert Poston on the front. It has the four continents and the female ostrich-feathered American Indian is seated on a rock. The gold banded maps of the world are of the outside latitude variety. An unmarked sheet iron false plate was used, but there is no marking of any sort on the moon or main dial. Another almost identical dial was made, presumably by the same person, for George Roberts of Birkenhead and Liverpool. The only date given for this clock maker is 1848, but one assumes he worked prior to this. The female American Indian in the corner of this dial is seated on a crouching Negro.

Luckily, Samuel Hunt, whose only date is given as 1845 in Loomes's *Dictionary*, used both cast and sheet iron false plates with his name on them. He was listed in *Gores Directories, Liverpool* first in 1821 as a

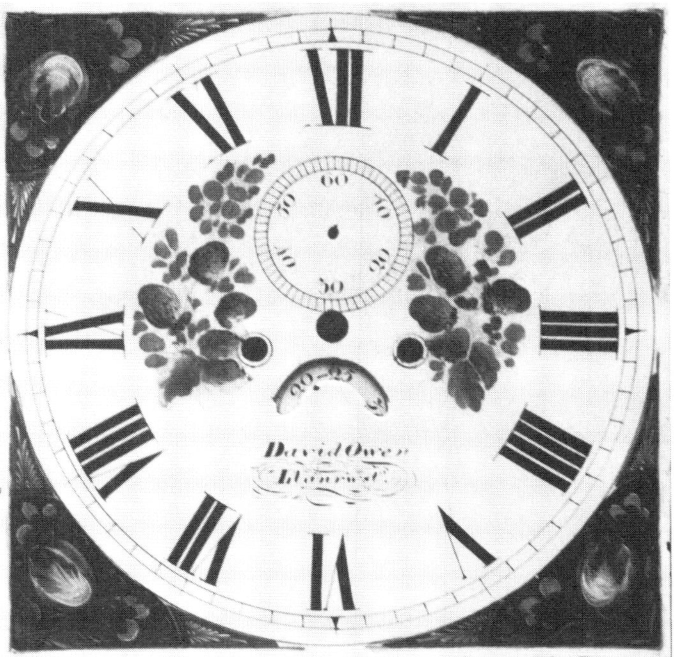

101 A Samuel Hunt of Liverpool made the dial for David Owen of Llanwrst *c.* 1845 with a cast iron named false plate (Illustration No. 4). The artwork is very boldly painted. See pages 155–6

Birmingham and Sheffield Warehouse, and later in 1823–9 as an Ironmonger. He probably sold Birmingham painted dials, and seems to have been influenced by those of E. Hobson *c.* 1830 because some of the Hunt dials seem remarkably similar to those of Hobson. His style was very colourful and vigorous, to say the least. One dial (**Illustration No. 101**) was made for David Owen of Llanwrst, whose dates are unknown, and it has the typical bright corners which in this case feature fancy tulips. The clock belongs to the National Trust in North Wales, and it has a cast iron false plate (**Illustration No. 4**). Another dial, donated by the author to Prescot Museum, was made for John Jones of Caernarvon (1828–44) (**Illustration No. 102**). Though, unfortunately, it had no false plate, the Owen one fitted exactly. The corners, using some gold leaf, are brilliant and very unusual. Another attractive dial by Samuel Hunt which had a cast iron false plate was made for Owen Owen of Amlwch, with tulip corners framed in pale blue and morning glories in the centre. The most spectacular and probably latest Hunt dial to be documented was one made for William Owen of Holyhead 1840–4 (**Illustration No. 103**) which used a named sheet iron false plate. The

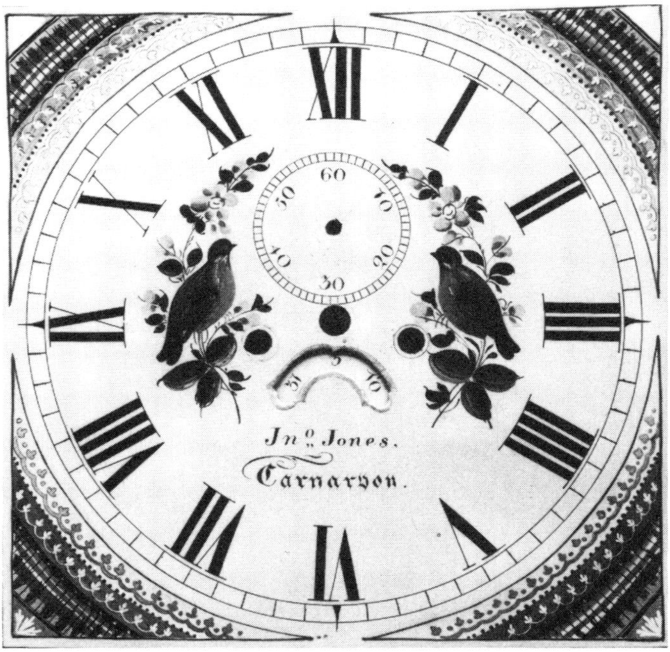

102 A Samuel Hunt dial made for John Jones of Carnarvon *c.* 1844. This dial, which has no movement attached, was presented by the author to the Prescot Museum in June 1992. The corners are very brightly painted. See page 156

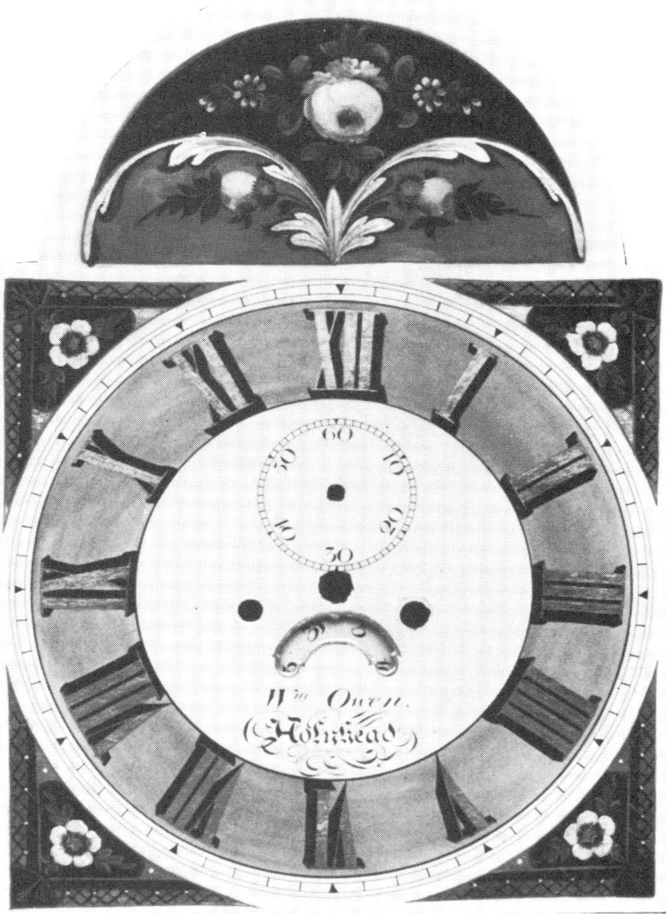

103 The most spectacular and latest of all the Hunt of Liverpool dials (1844–50). (See jacket illustration.) The mauve chapter ring, gold numerals with green tartan in the corners and a very bold break arch make this an exceptionally striking dial. The clock maker was William Owen of Holyhead and the false plate was stamped sheet iron. See pages 156–7

colouring is brilliant and the mauve chapter ring with its gold black shadowed numerals is most unusual. All of these Hunt dials were probably sent by sea, as at that time ships from Liverpool could easily reach all these ports. More of these Hunt dials do exist, but so far they seem to have escaped notice.

Another possible dial maker listed in the 1824 *Gores Directory* was William Lovatt of 57 Upper Islington, a miniature portrait painter and watch and clock dial enameller. One late dial maker of Liverpool was John Wilkes of 1 Upper Park Street, Toxteth, who was named as a dial maker in the street directories of 1865, though it is not known just what sort of dials he produced. A problem in trying to trace painted clock dial makers is that directories do not always differentiate between watch and clock dials. There was a Joseph Ball, 1845–65, listed as watch maker, watch dial maker and painter, who could well have made painted clock dials, and another firm, Cameron and Poston (probably Robert's son William), were listed as dial makers in 1849.

THE UNKNOWN DIALS AND RARE DIAL MAKERS

At the end of a survey such as this, there are bound to be the dial makers mentioned by others with no dials recorded, or very rarely seen dial makers, or dial makers whose place of work is not known. The earliest of these known to the author is a dial with a cast iron false plate with 'Beard & Co' on it. The dial has spandrel type corners almost identical to one made by Caleb Nicholas of Birmingham (*c.* 1785–90), which can be seen on p. 89 of Brian Loomes's *White Dial Clocks*. It also has a very decorated break arch done in gesso, and brass grommets were used. It was made for George or John Norton of Ipswich (1790–4), which seems to be about right for this dial, though no record of this company can be found.

There must have been quite a few unemployed local makers of moon dials for brass dialed clocks, once the fashion for painted dials took hold. The earlier moon dials had an engraved and silvered edge, but later on the edge was painted white, with black numerals. Such an unknown moon dial painter went ahead to make quite a large 14″ painted dial for William Parker of Burton, who is listed as working in nearby Tutbury in 1795. It was a break arch moon dial type, and the moon dial was made of iron. The main dial gives an effect of general lack of expertise. The ground paint is very thin indeed, and flat gold leaf is used to frame the floral corners. The arch over the moon dial is decorated in a most unusual way, and the Arabic numerals are not very crisp. The gilded winding holes are very close to the oversized date aperture and the iron date dial is held on by a stub foot and pin on the main dial. The whole effect reminds one of a brass dial, but work like this is probably rare.

Another 'unknown' is W.H. Price, B'ham, stamped on date dial and cast into the false plate. The author has not as yet seen this dial, but it was made for Samuel Maddock of Winsford whose only date is 1834 at Over Budworth. Samuel Maddock must have worked earlier than this, because the author has restored a dial made by Felton of Birmingham (1803–31) which had flared tumbling Arabic numerals and small quarters, also made for Samuel Maddock, probably *c.* 1815.

James Richardson of Birmingham made at least one dial with geometric corners, tumbling Arabic numerals and quarter numerals, with maps and a moon dial, but thus far there is no trace of this dial maker, who appears to have worked 1815–20. Baldwin and Munslow of Birmingham made a dial for Evans of Pwllheli, who died in 1820. There seems to be no trace of this firm either, but the cast iron false plate is very clear and sharp.

An article in *Clocks* magazine (June, 1991, pp. 21–5) illustrates a dial made by Cox of Taunton with a cast iron false plate. This firm has a date of *c.* 1810, but the dial is later than this with Roman numerals, Arabic quarters and a moon dial. There must have been maps in the hemisphere, but no trace of these can be seen in the large colour photographs. The clock makers were John and Joseph Thristle of Stogursey whose dates are rather vague as they were listed just working before 1840. The Arabic numerals are rather fat in the 1835–45 style, and the corner decorations are red ovals with a lot of gold leaf foliage. The moon dial was painted in a rather stiff style. It would be very interesting to know if other Cox dials exist.

Also in a more recent issue of *Clocks* magazine (August, 1992, pp. 44–5) Brian Loomes mentions false plates with the names of 'Wasborough Hale & Co., Bristol', on one and 'Langdale & Todd, Birmingham' on the other. Nothing definite seems to be known about either one of these makers, except that they must have existed as evidenced by their false plates. Another late dial maker was 'Neil' stamped on a sheet iron false plate, but it is not known what the dial was like. Brian Loomes's lists are full of names not only of dial makers but anyone associated with that trade. Though the author can only comment on dials or false plates that actually exist, there must be much more yet to discover about these 'unknowns'.

THE INFLUENCE OF EARLY BIRMINGHAM DIALS ON THE CLOCK DIALS OF THE UNITED STATES

The earliest American dials were brass and silvered ones with a few iron ones rather crudely painted, as well as some pewter dials. By 1730 moon dials began to appear, the earliest ones being engraved, but by at least 1760 many of the moon dials were painted. A number of clock makers emigrated from England to America, and one of the foremost of these was Thomas Harland, who left London in 1773 to settle in Norwich, Connecticut, where he set up a very profitable business in clocks and watches. His clocks mainly had one piece silvered dials, which he also supplied to the expanding clock trade, but by 1785 he purchased painted clock dials from England. These were used by many of the clock makers up and down the eastern seaboard of the United States. Ephraim Clark of Philadelphia, who started selling clocks as early as 1780, was

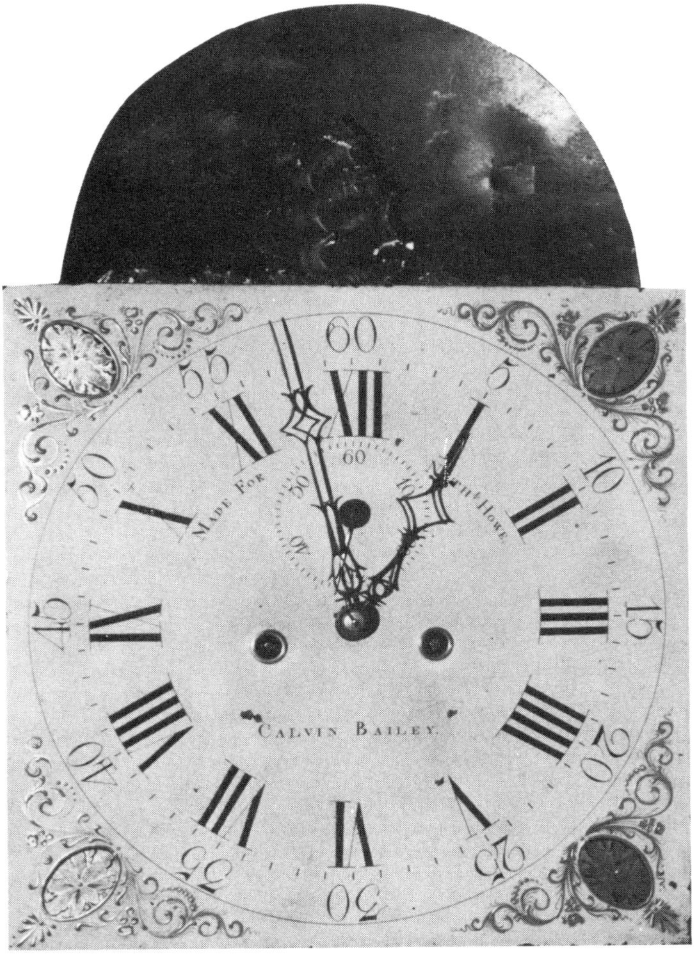

104 Osborne of Birmingham square dial with added break arch and automata. The dial is *c.* 1790 and the clock maker was Calvin Bailey of Hanover, Massachusetts. See page 160

importing cases of Birmingham painted dials from Liverpool by 1790. There were some enterprising clock makers using these Birmingham dials as early as 1775. The Birmingham makers that exported to the United States were Wilson, Osborne, Keeling and Ashwin and, after 1811, Walker and Hughes. As previously mentioned, these dials were exported without any clock maker's name on them. Both break arch moon and decorated break arch dials were popular, but square ones were also used. At least two of these square dials had an American addition to them consisting of a slightly dished break arch with a rocking ship automaton. This dial (**Illustration No. 104**), which can be seen at

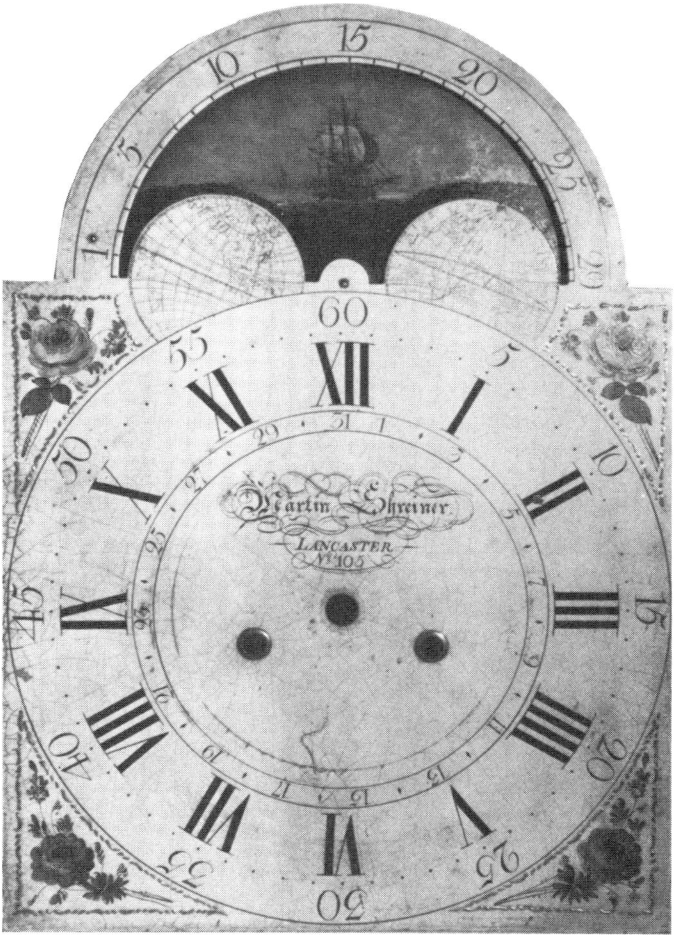

105 A Wilson of Birmingham dial that Martin Shreiner of Lancaster, Pennsylvania used in 1795. His signature is one of the most elaborate to be found on an imported dial of this type. Lancaster clocks favoured centre date hands. See pages 161 and 162

Old Sturbridge, Massachusetts, was made by Osborne for Calvin Bailey of Hanover, Massachusetts (1761–1828) about 1800. It was presumably the clock maker who did the modification and had his name as well as the customer's name put on. The presence of many English dials without any name at all on them has provided an irresistible temptation for the unscrupulous to add well-known early clock makers' names, but usually this is detectable. Some still exist in their blank form, including one Osborne dial with spandrel type corners and full Arabic numerals which is still in the author's family.

A dial where the clock maker could add his own name meant an opportunity to do as many flourishes as he liked. This Wilson dial

(Illustration No. 105) was signed *c.* 1795 by Martin Shreiner of Lancaster, Pennsylvania (1769–1866). It shows one of the most elaborate of American signatures ever seen, and it also gives the number of the clock: 105. In all other respects this is a standard Wilson dial, unrestored, and in a remarkable state of preservation in a country where painted dials seem to have deteriorated faster than on their home ground. Centre date hands were not all that common, but Wilson did produce quite a lot of dials with this type of date notation. (This dial was photographed at the Museum of the National Association of Watch and Clock collectors.)

At the same time as the Birmingham dials were being exported, there was a thriving business of dial making in the United States. These dials were of various types. The most ephemeral of these were the dials of painted paper, glued onto wood. Gideon Roberts of Bristol, Connecticut (1749–1813) often used this type for longcase clocks, some of which still exist, though the once white paper has darkened considerably with age. Paper on wooden dials was used as early as 1790. One Roberts dial *c.* 1800 was influenced by the spandrel type gesso corner decoration, and it had full Roman and Arabic numerals. Often the paper dials were coloured by hand.

Another paper dial with Wilson type flowers on it was used by Alexander T. Willard of Ashby, Massachusetts (1774–1850) about 1810. Imported dials carried a heavy tax, so there was considerable impetus to develop a dial making industry. What follows is a list of some of the known dial makers:

Christian Eby	Manheim, Pennsylvania	1793–1803
Benjamin Whitman	Reading, Pennsylvania	1799
Gideon Roberts	Bristol, Connecticut	1749–1813 (paper dials)
Willard & Nolen	Boston, Massachusetts	1805–9
Curtis & Nolen	Boston & Philadelphia	1809–17
Henry Nolen	Pittsburgh, Pennsylvania	*c.* 1817
Charles Bullard	Boston?	1816–71
William Jones	Philadelphia	
Patton & Jones	Philadelphia & Baltimore	1804–14
James Harrison	Waterbury, Connecticut	1804–6
Riley Whiting	Winchester, Connecticut	1785–1835

Not all the Birmingham dials were copied exactly or with any great professionalism, as can be seen by this example **(Illustration No. 106)** made for an unknown clock maker before 1800. (It was photographed at the American Clock & Watch Museum, Maple Street, Bristol, Connecticut.) The dial was constructed of cast iron, and the rather crude corner decoration featured quite a bit of red. The maps were

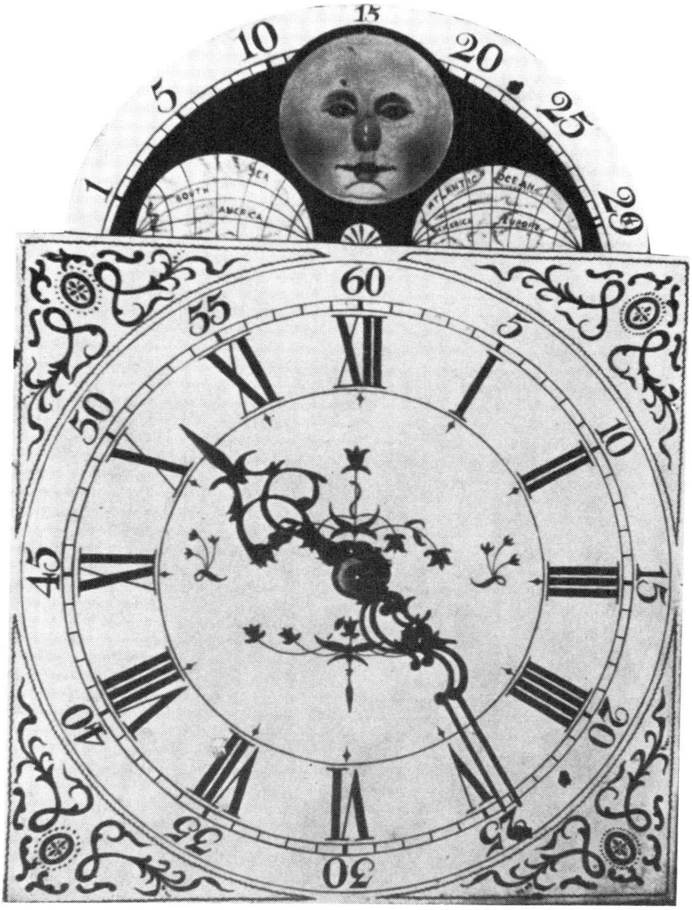

106 American-made rather crude dial with imitation moon dial. The clock maker and exact date are unknown. See page 162

rather odd, and the imitation moon would not have fooled anyone. The centre decoration was unlike anything known on an imported dial.

In a country as large as the United States there were obviously various centres for dial making. The earliest recorded dial maker could have been Christian Eby of Manheim, Pennsylvania (1793–1803) who was part of a flourishing clock business in Pennsylvania together with his two sons George and Jacob. Christian Eby produced at least one 30-hour dial which still exists. It was made *c.* 1800, with floral corners, framed in dot gesso, a floral break arch and Roman numerals plus Arabic quarters. The Birmingham influence was quite strong, though the layout was subtly different, and the artwork was bolder and more stylized. Interestingly enough, one of Christian's sons, Jacob, used an Osborne dial with gesso corners, maps, moon dial and centre date markings

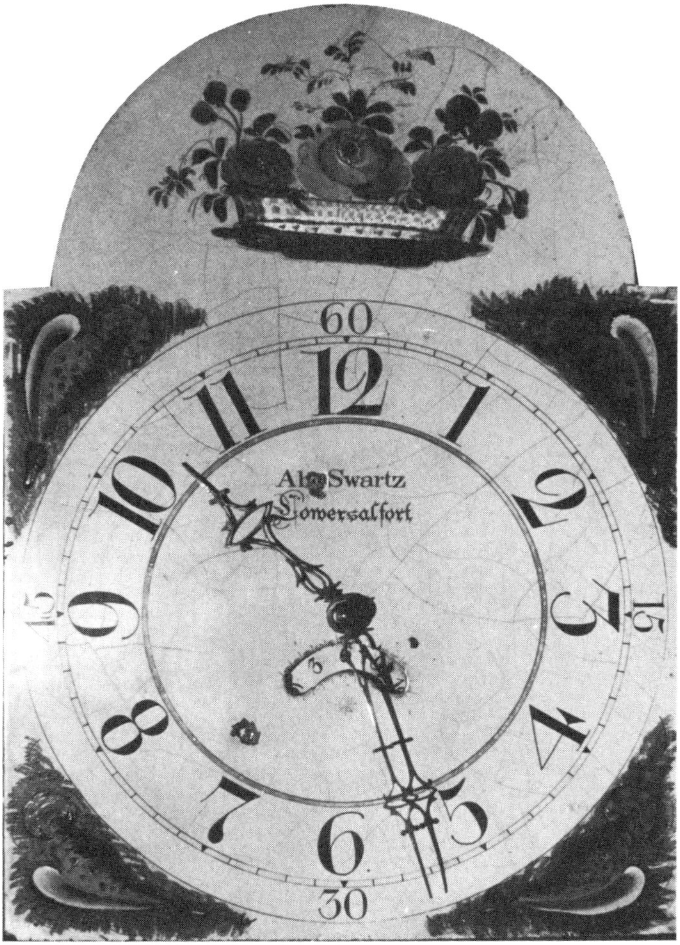

107 Pennsylvania 'Dutch' dial made by Abraham Swartz [sic] possibly *c.* 1815. See page 165

c. 1805. The use of the centre date hand was popular in Pennsylvania.

Benjamin Witman or Whitman of Reading, Pennsylvania (1774–1857) also made some of his own dials, one of which still exists. It was a 30-hour clock dial *c.* 1801 but is in very poor condition due to flaking. This dial has the conventional Birmingham layout with Roman numerals and Arabic quarters, decorated date aperture and Birmingham type floral corners, but the break arch contains an eagle with a shield. Lancaster county in Pennsylvania was particularly rich in both clock and dial makers, and it was also the centre for a number of German settlers so that the Black Forest influence can be seen on some of the dials. Chester and Berks county also had a lot of clock makers, but it is not known if any of them made their own dials.

The earlier pre-1810 Pennsylvania dials followed the full Roman and Arabic numerals, but they gradually went over to Arabic numerals only, such as this one photographed at the American Clock and Watch Museum (**Illustration No. 107**), made for or by Abraham Swartz of Lower Salfort. The Arabic numerals were often very large, and as time progressed, the corner decoration became bolder. Sometimes false plates were used, but generally these were unmarked with any maker's name. The 30-hour clock was very popular in the Pennsylvania area. Various types of maps were used for those dials with moons, and some of these from about 1800–20 had outside latitude numerals on them, though others did not. The maps appear to be transfer prints, yet a few were as crude as those used on Manchester made dials. It is difficult to know when this area ceased to produce longcase clocks with their dials, but a guess would be that around 1830–40, when the cheaper type of shelf clock became more popular, the longcase clock making faded out. Some of these shelf clocks were made locally, and could have had local dials. Many of the dials produced in this region were decorated with shells in the corners similar to the Swartz clock dial, and this motif was also used on the bracket clocks.

The Boston area certainly was a dial making centre, with a great number of clock makers to supply. Most of the American dials made for Aaron and Simon Willard, William Cummens and Elnathan Taber were made by Charles Bullard (1816–71), but there must have been dial makers around before that date. Benjamin Willard of Grafton, Lexington and Roxbury, Massachusetts, probably used a local dial maker for at least one of his clocks (**Illustration No. 108**), where the fine gesso work is up to Birmingham standards but the maps definitely are not, as they look hand drawn. The shape of the date aperture is also unusual, as are the black edged winding holes. This was photographed at the American Clock & Watch Museum, Bristol, Connecticut. Unusual date apertures also appear on at least two other presumably Boston-made dials *c.* 1795. One dial, also photographed at the American Clock & Watch Museum, was made for Joseph Chadwick (1787–1831) of Boscawen, New Hampshire (**Illustration No. 109**). The corners are direct copies of Wilson type flowers, complete with gilded gesso framing, but the break arch is much more elaborate than was generally found on Wilson dials. The date aperture is tiny, with square ends. Another dial with the same type of date aperture was made for Simon Jocelin (1746–1823) of New Haven, Connecticut. The corners of this dial were of the fan type, surrounded by a frame of straight gilded gesso, and the break arch had a bunch of Wilson type flowers. These Wilson type flowers appear in the corners, surrounded by a frame of straight gilded gesso in another dial made for Jacob Alrichs of Wilmington, Delaware *c.* 1800, but the graphics have gone over to the

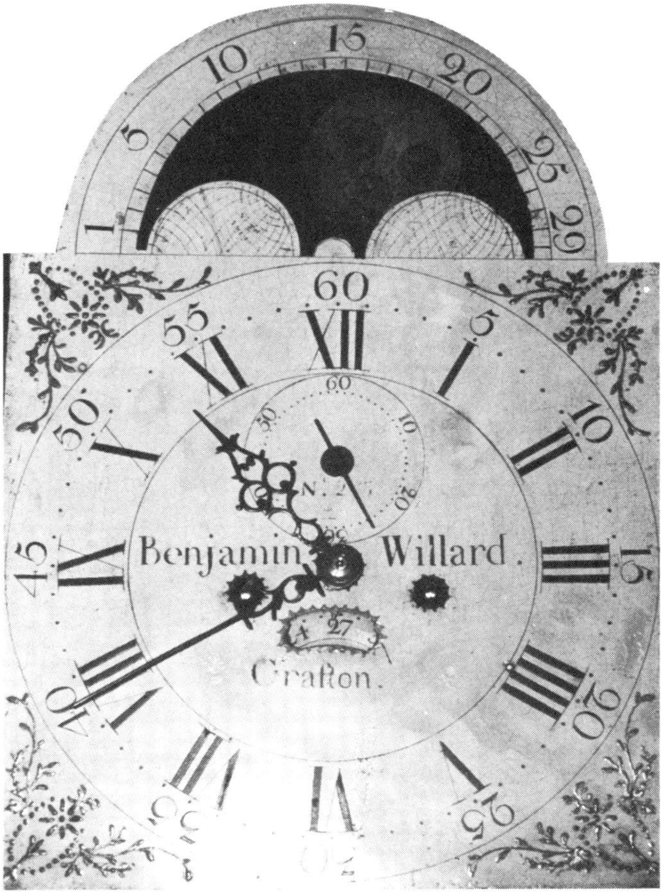

108 A Benjamin Willard dial made locally to Grafton Roxbury and Lexington, Massachusetts. The gilded gesso work has been influenced by imported Birmingham dials. The maps are hand drawn. See page 165

newer style of all Arabics. The fan corner, first seen on Wilson and Osborne dials, continued to be popular in America for quite some time.

Probably the best known firm of American dial makers was Curtis of Boston. There were several Curtis brothers, all of them artistically talented. Samuel Curtis (1785–1879) had been a partner with his younger brother Benjamin B. Curtis, the ornamental and dial painter (1794–?), for a short while, though Benjamin later became a partner with Charles Hubbard. The Curtis firm really started off as Aaron Willard Jr. and Spencer Nolen *c.* 1805, using cast iron false plates marked Willard and Nolen. This partnership did not last long, and in 1809 the business became 'Curtis and Nolen' of both Boston and

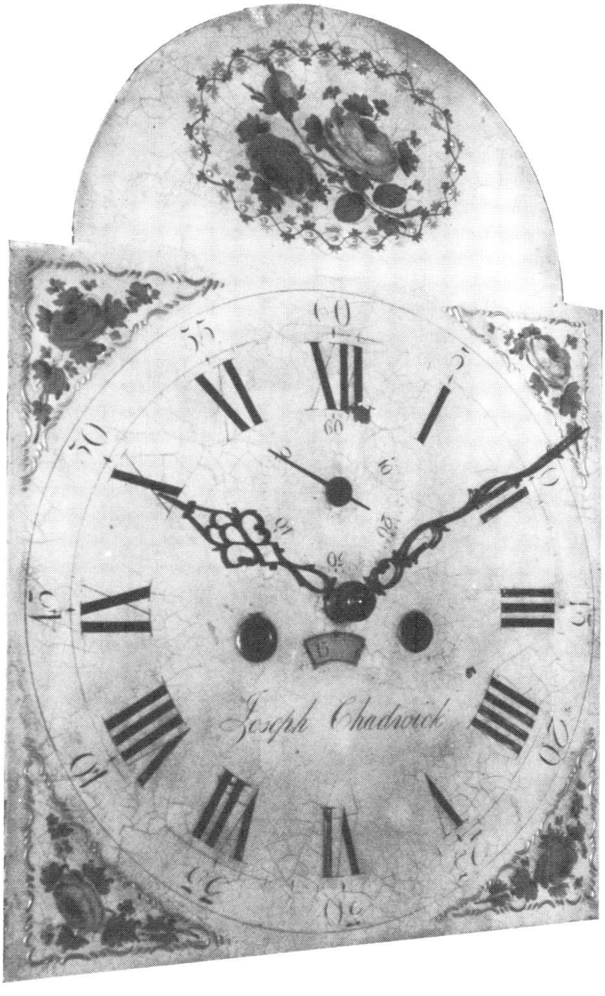

109 Boston-made dial showing Wilson influence on the corners. The date aperture is very small and it was used by Joseph Chadwick of Boscawen, New Hampshire *c.* 1795. See page 165

Philadelphia. The firm lasted until about 1820. During the latter period of the Curtis and Nolen partnership, Benjamin B. Curtis established himself as an 'ornamental painter' near the Curtis and Nolen building, and in 1820 Samuel joined him. Soon after this Charles Hubbard became a partner, so the firm was then called Curtis and Hubbard until 1824, when Samuel Curtis resumed business alone until his death in 1879. The Curtis firm not only made dials but did all sorts of ornamental painting, gilding and mirror making. Charles Hubbard also worked separately after 1824 and an advertisement he placed in the

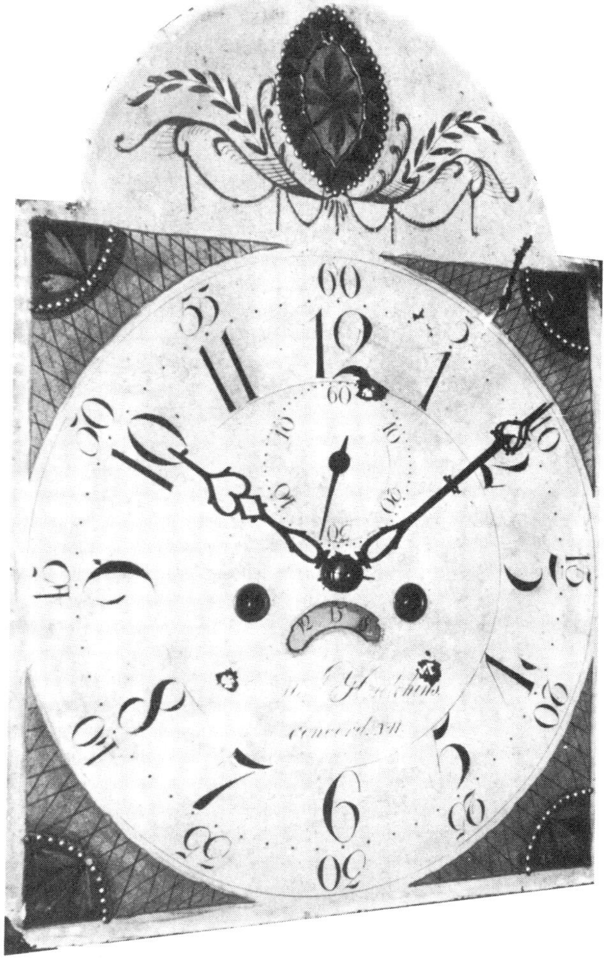

110 Samuel Curtis made this dial for Levi Hutchins of Concord, New Hampshire *c.* 1800. Note the use of double Arabic numerals and dotted gesso work. See page 168

Boston Directory states that he did 'Sign, Block, Ornamental and Military' painting with 'a good assortment of timepiece dials'. Curtis dials were always of good quality, on a wrought iron base, and were well executed.

In his earlier work Samuel Curtis used Arabic numerals, such as this one (**Illustration No. 110**) made for Levi Hutchins of Concord, New Hampshire *c.* 1800. The numerals, though bold, have great style, and the corners together with the break arch are artistically pleasing. The dot gilded gesso seen here is typical of Curtis and was often used. Later

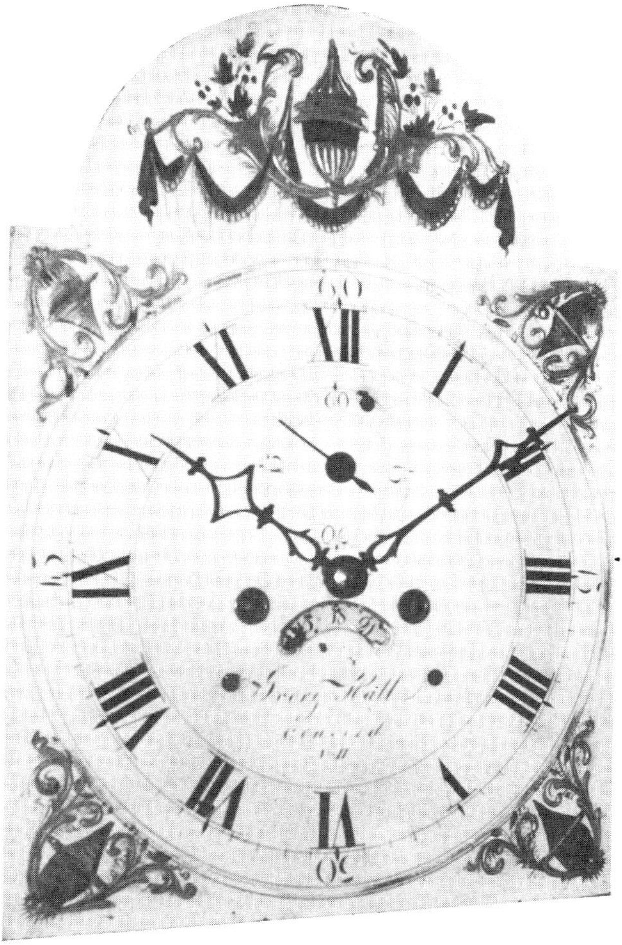

111 Curtis-made dial for Gregory Hall of Concord, New Hampshire *c.* 1815. This dial features elaborate gesso work and a return to Roman numerals with Arabic quarter numerals. See pages 168 and 169

dials went over to the Roman numerals with small Arabic quarters such as seen here, photographed at the National American Clock & Watch Museum, Columbia, Pennsylvania (**Illustration No. 111**). It was made for Gregory(?) Hall, also of Concord, probably about 1815. The dot minute marking has changed to the barred circle, and the whole chapter ring appears smaller than in the Hutchins dial. Curtis quite often used the gesso shield motif, usually with stripes on it, though sometimes these shields were plain. The corners and break arch decorations in the Hall dial appear to be a cross between a shield and an urn. The shield was used on the earlier type of dial, with full Roman and Arabic numerals as

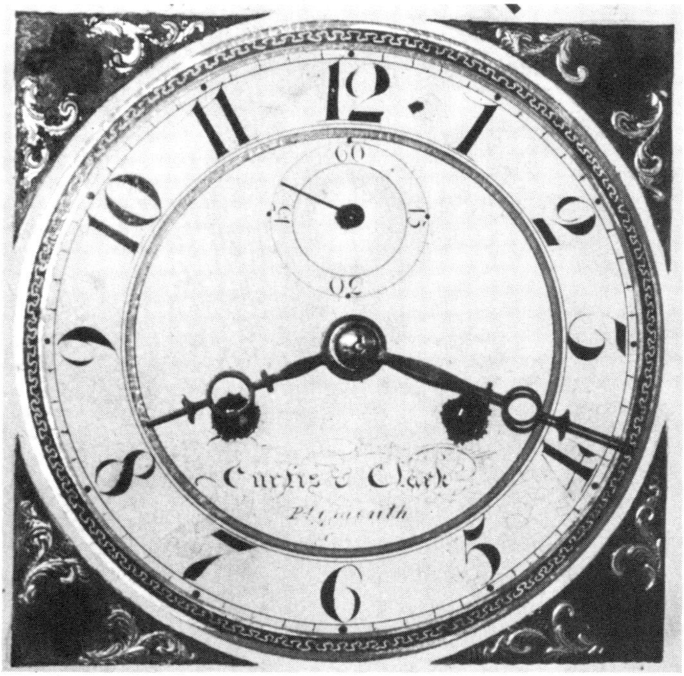

112 Curtis-made shelf clock dial *c.* 1825 used by Curtis and Clark of Plymouth, Connecticut. This dial is in rather poor condition due to flaking paint. See pages 169 and 170

well as the later variety. The elaborate gesso work became simpler as the century progressed.

Samuel Curtis also made a number of dials for shelf clocks, wall clocks of the Banjo or Girandole type, as well as dials for miniature wall clocks. Some of his dials were of the dished or convex type. This shelf clock dial (photographed at the National American Clock & Watch Museum, Columbia, Pennsylvania) (**Illustration No. 112**) was made for Curtis and Clark of Plymouth, Connecticut in 1825, and it is typical of his work at that period.

The small quarter markings have gone, but there is still quite a lot of gilded gesso work in the corners of a green background. The 'name' has a different type of lettering, surrounded by a lot of scrolling. No grommets appear to have been used.

Quite a few of the American clock makers appear to have made their own dials, with varying results. Because of the long distances involved and problems of transportation, the author has not restored many American-made dials, but this one was in such very bad condition that the owner was willing to wait several years for the work to be done. This

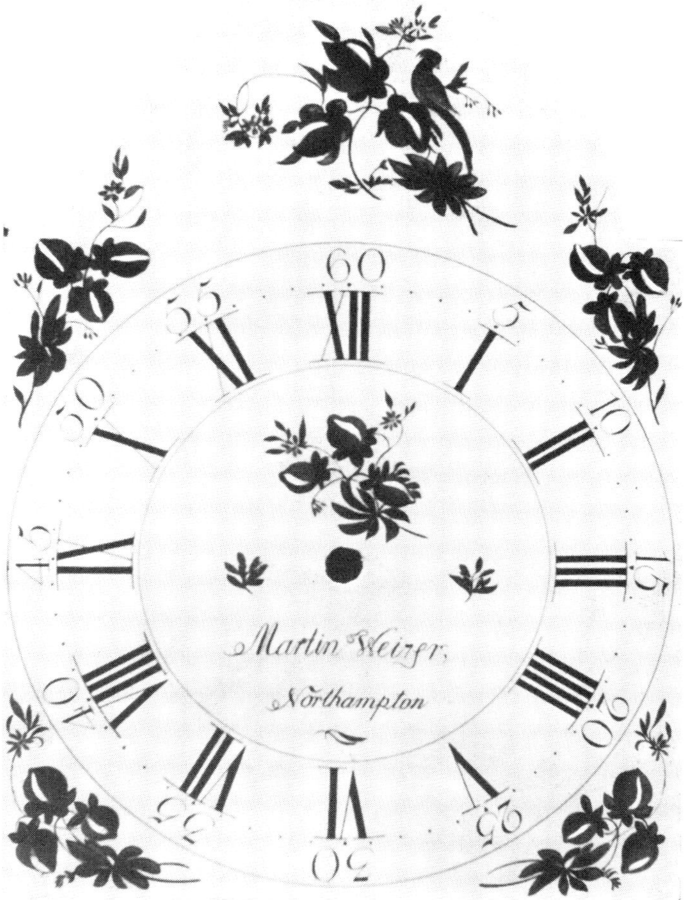

113 Dial probably made by the clock maker Martin Weizer of Northampton, Massachusetts *c.* 1800, with very bold and colourful artwork based on a thin iron plate. See pages 170 and 171

dial (**Illustration No. 113**) was made by Martin Weitzer of Northampton, Massachusetts about 1800. The iron sheet was very thin, and the off-white ground paint was not thick enough to prevent the red of the iron oxide priming coat showing through in places. There also were large areas of cupping and paint loss. The graphics are standard Birmingham style as is the layout, and though the decoration could have been influenced by English dials, the manner of painting is far more vigorous and bold with considerable use of bright red.

Contemporary with Samuel Curtis was the dial maker William Jones of Philadelphia, whose exact dates are unknown. He was said to have produced very fine dials, 'baking them in an oven after painting in order to produce a glossy finish'. He could not continue after 1840 because his hands became paralysed. Abraham Patton and Samuel G. Jones of Philadelphia and Baltimore made painted iron dials from 1804–14. This firm had a named cast iron false plate, one of which they supplied for a clock with a moon dial made for Richard Jackson of East Springfield, Ohio about 1828, and either the dial had sat around for a long time or Patton and Jones continued making dials into the 1820s.

Gradually the relatively expensive longcase clock was phased out in favour of the cheaper and eventually mass produced shelf clock, as happened in Pennsylvania. The wrought iron dial was supplanted in the shelf clock by the thin zinc dial, which could easily be cut and pressed to give a raised chapter ring effect. The decoration on these zinc dials consisted of a rudimentary flower pattern done with a stencil which was a far cry from those dials inspired by Birmingham. As a matter of interest, these zinc dials used no priming coat, and even the base paint was very thin, not adhering well to the smooth surface of the metal so that considerable flaking was liable to happen as the dial aged.

Metal was always expensive and often in short supply in the early days of American clock making – so much so that whole movements were made of wood, and painted wooden dials became quite popular after 1800. Eli Terry (1772–1853) went over to using these wooden dials in 1806. The earliest versions of the painted wooden dial had corners quite reminiscent of the Birmingham dials with dot and dash gilded gesso corner framing sprays of flowers, though the flowers were done in a distinctive style with little detail. This type of corner decoration, together with quite bold straight set Arabic numerals and small quarter Arabic numerals inside the chapter ring, was used by Eli Terry as late as 1818 for a shelf clock of the pillar and scroll type. Thomas and Hoadley of Plymouth, Connecticut also used painted wooden dials on a longcase clock *c.* 1812, and this dial had full Roman and Arabic numerals. As late as 1825 Silas Hoadley used a wooden painted dial with gesso framed corners, tumbling Arabic numerals and an urn in the break arch for another longcase clock. There were two false winding holes painted on the dial, though of course the clock was a 30-hour one.

Riley Whiting of Winchester, Connecticut (1785–1835) produced a number of painted wooden dials for longcase clocks that have been instantly recognizable because he always put his name on them (**Illustration No. 114**). This dial, which the author has restored, is a typical Whiting dial of about 1820, and the corners show the modified fan type of decoration neatly executed in gesso. The clock was a 30-hour longcase one, but false winding holes were done in paint and gold leaf. Certainly as late as 1830 wooden dials such as this one (**Illustration No. 115**) were used for shelf clocks and were often incorporated into a

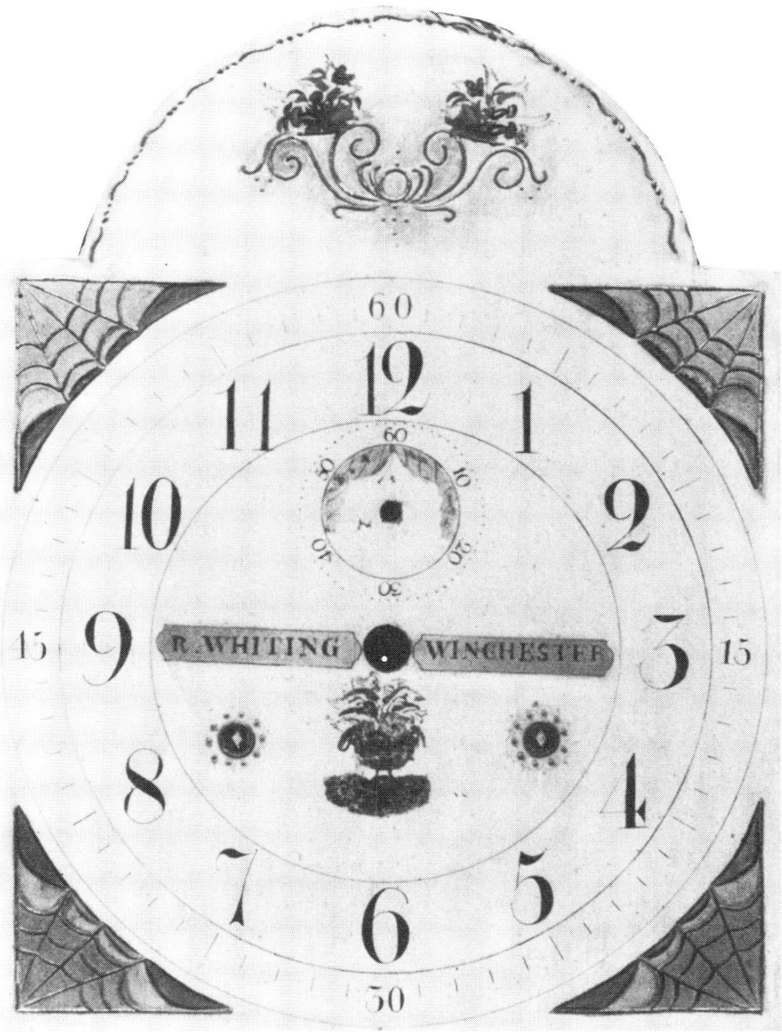

114 Riley Whiting of Winchester, Connecticut, made this wooden dial *c.* 1820 with mock winding holes and some raised gesso work. See page 172

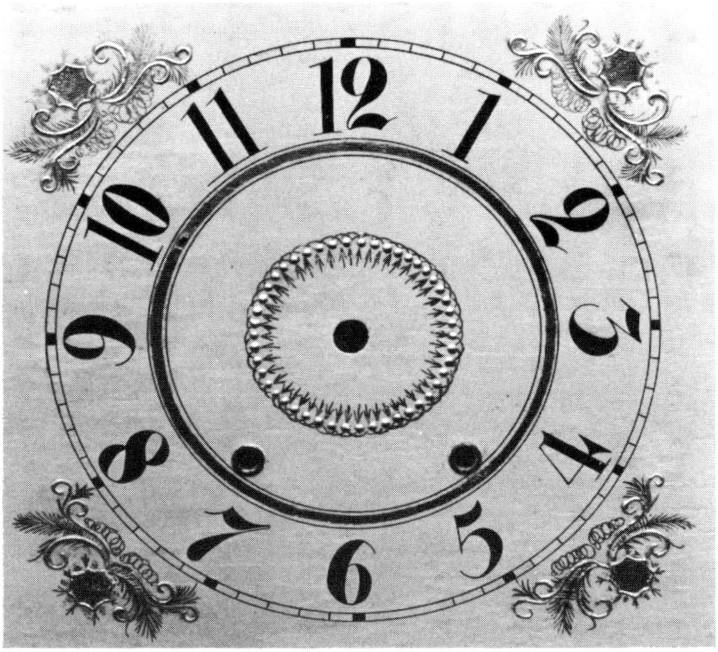

115 A typical wooden dial for a shelf clock *c.* 1830 with some raised work, in quite an elaborate pattern for the corners. See pages 172 and 174

clock where the main decorative interest was a reverse painted glass door panel. This particular dial had no maker's name on it, for that information was on a paper label, often elaborately printed and illustrated, inside the clock case. Gold leaf and gesso was still being used, though the base paint often seems to have suffered from transverse cracking following the grain of the wood which can be seen even after careful restoration. The wooden dial also gave way to the more easily produced zinc one.

The subject of painted dials in the United States is a very large one, and this section could not possibly deal with it all in detail, but the author has attempted to show some of the development that took place in a country where the painted dial has always been highly regarded.

5 Practical Dial Work

HOW TO LOOK AT OR OTHERWISE EXAMINE A PAINTED DIAL

Painted dials are often seen but seldom really noticed by the general public. However, it is quite different for the professional or student. Dials may be in a dim shop, poorly lighted gallery or stately home, but good light for examination is essential. Bright natural light is best, but in case of absolute need a portable light is better than nothing. A small hand mirror is useful when looking at the back of the dial to find the name on a false plate, moon or date dial. A photographic puffer brush, a magnifier or jeweller's eye glass, a soft cloth and a good pair of pliers help in examination, but often access is restricted to what is visible from the front only. The first thing to examine is the dial as a whole. This is far more easily done if the dial is allowed to be detached from the movement. A short resumé of this process may be useful.

First, put the key on the winding square. Let down the lines by pushing a very thin piece of steel between the click and the barrel teeth. Remove the weights and unhook the pendulum, being very careful movement and dial do not fall out of the case. A block of wood of appropriate thickness (about 2'') should be placed under the back of the backcock to hold the dial roughly horizontal when it is placed on a table, which should be protected with a cloth or layer of newspapers. Then the pin holding on the hand collets can be removed if deemed necessary so that the minute and hour hands can be taken off. The second hand should be carefully levered up. It is usually best to remove the bell, which is more fragile than most people realize. Pliers should be used to remove the pins that hold the false plate pillars to the front plate of the movement, or, in the case of no false plate at all, the long feet will have been pinned directly into the front plate. It goes without saying that everything which came off should be put back, including the old nails, split pins, dressmakers' pins and carpet tacks or bits of wire that have been used instead of proper clock makers' pins.

The appearance of the dial is the most important thing. Does it look right, i.e. does it fit into any of the main categories in the first part of this book? If it does not, then there is usually something wrong somewhere, and one may be dealing with a complete repaint, possibly done some time ago. These repaints are more common than people think and are sometimes hard to detect. The main thing that can cause a dial to look all wrong is the overpainting of the graphics, which is easy to detect if done crudely but much more difficult to see if well done. The lack of things such as flower stalks etc. can often be seen if the dial is tilted so that the light comes in at an angle. This is the restorer's favourite 'raking light'. Even very scrubbed dials often reveal names and even maps by this method. The author has never seen a dial where the hemisphere on a break arch moon type of dial was meant to be completely blank. Irrelevant notations on the front of a dial are all too easy to spot, such as 'Presented to Joe Bloggs on his Retirement, September 1929' or 'Restored by James Smith 1936'. Much more subtle and luckily quite unique were the scratched notations of the dates when a clock was repaired, done in the main numerals of the dial in no less than four places.

Next it is advisable to look at the condition of the dial. Is it just plain dirty, or is it rusty, damaged or flaking off from the metal? The reverse of a dial is often as important as the front, because it is usually full of information. The first thing is to try and find out who made it. Cast iron false plates frequently have a name, sometimes two names on them, but there are unmarked ones, or false plates with very small numbers on them, impressed into the metal. Sometimes the false plate was attached to the dial the 'wrong' way around so that the name is next to the dial and very difficult to see. This is where the mirror comes in useful. The moon and date dial often have a name stamped on them if one could but see through the dirt and often heavy corrosion. Dust alone can be removed by a cloth, being careful not to catch the sometimes fragile teeth in the cloth, but anything else will not come off. The name on the false plate should correspond with that on the moon or date dials, though there have been some cases where, though original, they do not match. This is usually more common with date dials. Some dial makers stamped their dial plates on the back, but this can be quite faint or clogged with paint. Wilson often used a Maltese Cross, but others put only initials, 'W. & S.' for Whitaker and Shreve, for example. Presuming the name of the dial maker can be found and there is a clock maker's name on the front of the dial, a quick check in various standard reference books together with a list of dial makers and their dates would corroborate that the clock maker could have used that particular dial maker. However, the dictionaries or other reference books cannot be one hundred percent accurate, and many clock makers are not included at all.

The back of the dial can show any major alteration to the dial, such as the cutting off of the original feet because they did not fit a replacement movement and therefore substitute feet had to be put on in a different place. However, when this has happened the resultant holes and filling can be seen on the dial front. Even winding holes can be altered, but these usually look all 'wrong' on the front of the dial. Date and seconds holes that have been moved can also be seen on the back, but some original dials had the holes for date or second hand plugged with lead and a decoration painted on the front at the time of manufacture. It can also be seen from the back if the winding holes, date or seconds holes have been made before painting, as the pale paint of the front runs through and shows up on the darker paint on the back. This could be considered to be a minor point. Also, the fingerprints of the dial worker in pale base paint are often seen, as well as paint can marks. Drips of candle grease are quite common, and any label or remains of a Wilson job ticket should be examined in detail.

Next, and far more difficult, is the detection of scratched or painted-on marks. This is absolutely impossible to do without detaching the dial or even the false plate from the movement. Some marks can be hidden under the moon dial. In the case of the Wilson dial painter's mark done in black, this is usually found on or near to a white splodge of paint. Other firms, such as Byrne and Whitacre and Shreve, may have put on quite large numbers which are often obscured by dirt and grease. These numbers, usually done in white base paint, can be made more visible by removing some of the dirt and grease using a grease solvent such as 'Genclene' (111-Trichloroethane) on a soft cloth. It is advisable to use only a very small quantity of the solvent and to avoid breathing in the fumes. The scratched-in markings are usually repairers' marks, often initials and a date, though sometimes 'clean'd' or 'lines' show what has been done in the past. It is often astonishing what can be found on the back of a dial, so one can be quite dismayed on examining a large collection of painted dials in a museum to find most of them have had a coat of thick black paint all over the back, thus obliterating forever any information that had been there.

The false plate also should be examined as some of these carry numbers on them which should correspond to any found on the main dial or moon. Cast iron false plates often seem to have been dropped, and the resulting breaks were usually mended in an easily visible way. The sheet iron false plates can be rather faintly stamped with a maker's name, though many are completely blank.

Date and moon dials when made of brass are often heavily corroded, and iron ones often are rusty. Sometimes there are bent or missing teeth. Often the evidence of stress cracking can be seen on the front of date or moon dials which show very fine cracks, sometimes with blue green corrosion coming through them. Some moon dials are fragile in the extreme, and almost disintegrate at a touch, so obviously these must remain in situ until, hopefully, a detailed examination and photographic record can be done. Moon dials usually have been thickly varnished when they were made, and if they have never been cleaned this varnish has aged to give a yellowed or even brown appearance.

Presumably by the time all this has gone on it should be clear just what the dial is, and its approximate age, as well as its condition in a superficial way. It can be very frustrating to the restorer – conservator to put back a dial obviously in desperate need of attention, but at least a written record will have been made.

DIAL AND FALSE PLATE PHOTOGRAPHY

Preliminary examination is one thing, but detailed recording of a painted dial is another. Obviously a written record should be made as well as the measurements, but in ideal cases a good photographic survey is invaluable. Endless curatorial time is wasted by the perusal of really bad and indistinct photographs, and these can appear in many specialized magazines as well. The fuzzy photograph also turns up in some books, though these tend to be the earlier ones. Good photographs are an asset in all sorts of ways, especially if a clock is stolen or damaged. Many museums keep good photographic records, as do most houses or castles open to the public, but there definitely are gaps in places.

Dial photography can be roughly of two types: the perambulatory, such as when visiting a museum, gallery, sale room, shop or stately home where the item cannot be moved, or sometimes not even have the hood door opened, and the studio work for serious dial recording. Many museums and stately homes will allow photography, though it is usually advisable to ask for this in advance, preferably in writing. Often flash is not allowed and sometimes tripods are barred. A good quality pocket camera or, ideally, a single lens reflex can be used. Kodachrome 64 has proved to be satisfactory for colour work. A cable release is almost essential, as is a sturdy tripod when allowed.

One of the main problems of this perambulatory type of photography is getting rid of reflections. It is much more difficult to photograph a dial with the hood door shut, but often the whole case is locked so as to discourage meddlers, and one just has to put up with it. If the flash gun is attached to the camera, one can only take the photograph slightly to one side to avoid the reflections. Ideally, the flash gun should be separate and an extension lead used with an assistant holding the gun to one side in order that the photograph can be taken head on. If a time exposure is necessary, a unipod is useful or else the camera can be propped up on the assistant's shoulder. In general, colour transparencies are better for recording purposes as they can easily be enlarged and the colour rendition seems more stable. Polaroid shots lack detail and tend to fade, but they can be used in emergencies.

Studio photography means one has complete control over the dial and background. If there are to be a series of photographs of different dials taken at various times or the same dial in various stages, it is important that the conditions remain the same. Natural diffuse light seems to give the best results. All of the photographs in this book, with the exception of most of the American section, were taken in a conservatory or greenhouse with a colourless plastic roof, done before the sun got around to cause problems. As seen in **Illustration No. 116**, a black background is used on top of a table with the dial leaning against it. There is a transparent plastic guard at the top of the background in order to prevent condensation drips in wintertime. A black cloth covers the table so there are no reflections. It is essential to adjust the camera so that the focal plane is parallel to the dial. There is a piece of mirror mounted on a board which has a T piece on the bottom, and an adjustable rod at the top to hold the board away from the dial. It is easy to adjust the mirror so that it is parallel with the dial in all directions. Then it is only necessary to adjust the camera so that the image of its lens, as reflected by the mirror, is in the centre of the view finder. The optical axis will then be normal to the dial, i.e. the focal plane will be parallel to the dial. Both stand and mounted mirror are not available commercially but can easily be made. The paint on the stand should be matt black.

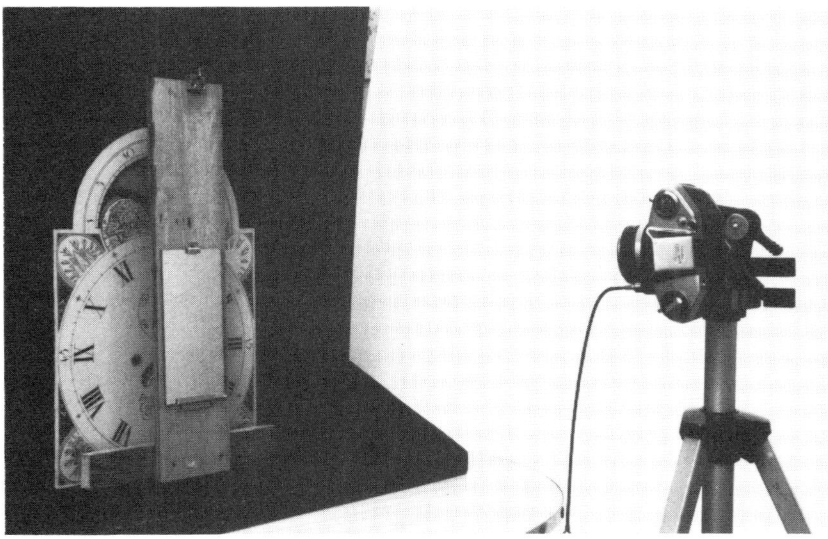

116 A photographic set-up using a mirror in order to get a square picture that is not distorted. See page 179

For colour work a Weston meter IV fitted with an intvercone for indirect light readings has proved satisfactory, but there are several other makes which will work as well for monochrome. The meter should be used as an integrating meter. Whatever method of exposure estimation is used, it is best to bracket the exposures taking three shots per dial, as often the dial is only available for a limited time. Colour slides should be stored in dust free slide boxes in an area that remains at as constant a temperature as possible, which is not too dry or damp. The best of the 'spares' should be kept separately in case of damage or loss to the original one. The problem of cataloguing slides is a difficult one, especially when the slide collection runs to thousands, but each restorer/conservator will have his or her own system.

Even in ideal situations, reflections, especially with a restored dial, tend to show up. The cure for this seems to be a large sheet of hardboard or plywood 3′ x 3′ painted matt black on one side with a hole in the middle for the camera. The light meter must be used with this shield in position, so it is definitely a two person job. When photographing a brass dial the shield should be reversed in order to avoid the black surface being reflected by the brass and silvering. With this sort of set up there is the possibility that the top of a break arch dial will be more brightly illuminated than the bottom. This is not so noticeable in colour slides but it does create problems with black and white work. When printing it is worthwhile doing a test strip for both top and bottom of the dial and then dodge it when making the final print, i.e. give more exposure to the top than to the bottom of the print.

There are some dials where the reverse is very interesting, for example an unusual moon mechanism or the use of rollers for a date ring. As the back of the dial is dark a white background should be used, and of course the shield is unnecessary. False plates need recording as well, though these can be very hard to photograph in their original state as the casting is often poor and the name indistinct. One solution to this is either to paint in the lettering with white water colour or poster paint. A more satisfactory solution is to dust the whole false plate with fine talcum powder which shows up the lettering to very good advantage. Needless to say, paint or powder should be removed as soon as possible after the photograph has proved satisfactory.

Moon dials are not usually detached and photographed separately, but they should be. Especially in the case of very late dials with no dial maker's mark, the moon is a very important aid to grouping dials together. Also, some very few moon dials are real works of art, or amusing commentaries on daily life, where it is necessary to see the whole moon dial and not just a part. Occasionally it is necessary to photograph the reverse of a moon or date dial, for example in order to show interesting marks, and these should be treated as brass dials.

Two types of dials can produce difficulties. One is the dial with a large amount of ornamental gilded gesso work which can almost disappear if the lighting and exposure is wrong. The other type has either gold leaf corners, as in the Halifax Newcastle dials, or bronze powder backgrounds used in 1840–60 dials. Both of these types come out dark in the black and white photograph so that it is hard to see the decorative elements. Filters have been tried but they are not much help. Occasionally detailed photographs are necessary, and this means using extension tubes or supplementary lenses.

Very rarely it is considered important to photograph a dial under UV light, either before or during restoration. Photographing the fluorescent image shown up by UV illumination is quite difficult. There is a lot of UV light reflected from the off-white background paint of the dial, and the eye does not see this as it is beyond the visible spectrum. However, the panchromatic emulsion of a film is sensitive to this wavelength and it will swamp the fluorescent image. It is necessary to cut out all the UV light entering the camera, and the only effective way to do this is to use a tri-colour red filter. Also, one has presumably just a single UV lamp which has to be at one side of the dial so that the illumination is very uneven. However, this can be corrected during printing by giving greater exposure to the darker parts of the image to a factor of two or three times.

Another valuable asset to recording a dial is the use of pencil and tracing paper where details or signatures appear fugitive or suspect. In the case of very extensive restoration, such as can be found with rust damage for example, the whole dial should be traced as well as photographed before any work is done on it. Where the dial has a verbal history from the owner this should be briefly written down as it can provide valuable clues. An example of this was a dial where the numerals were heavily overpainted with a type of very thick black/brown paint not previously encountered. It was discovered that the clock had been owned by a bargee who found tar paint a useful touch up for his dial.

REASONS FOR CONSERVATION AND RESTORATION OF DIALS

There is a school of thought that holds it is better to leave a dial untouched than attempt to do anything at all. This idea is soon disposed of when one has seen, in a well known museum, a working longcase clock with an unstable flaking paint surface to the moon dial busily shaking off this paint with every hour's strike. Each dial is unique, and they are not static items but change over the years.

Ignorance and accidents take a terrible toll in addition to malpractice in restoration, hazards in handling, vandalism and inherent problems because of unstable paint or varnish. No schools of dial restoration exist as yet in this country, and there are no standards which apply to this type of work. Exactly the same was true of the fine art restorer a long time ago, and the ravages of injudicious past work are still being tackled today. One can not do better than quote from p. 63 of the Murray Pease Report concerning the obligations to respect the integrity of the object.

Thus having put, hopefully, a convincing case for conservation and restoration of a painted dial, the advantages of such a course of action are as follows:

1. Prevent further damage or deterioration.
2. Remedy any dangerous conditions and remove spurious additions.
3. Provide accurate information about the dial.
4. Give, in the case of a museum or house open to the public, an accurate idea of what the dial originally looked like, allowing due credit for age, wear and tear.

The decision to have work done on a painted dial is not one to be taken lightly, either by a public institution or a private owner. It should be done in conjunction with any necessary work on the movement or case, as it is quite useless to restore a dial from an old clock, replace it, and then expect the dirty worn mechanism to carry on as before. Also, it is essential to check that the moon and date dial, if present, will work properly. The author has known of several instances when returning a restored dial together with its cleaned mechanism that the owner has been so pleased with the result the case was then deemed worthy of attention.

Choosing a conservator/restorer is extremely important. One that is on the list of accredited firms of the Conservation Unit of the Museums and Galleries Commission should be satisfactory, but ideally the premises should be visited and previous work examined. A restored dial ought to look clean and attractive but still show its age; i.e. the craquelure which is usually present should be visible. It is most inadvisable to turn over a dial to an antique dealer who 'knows someone and will touch it up' because there is no control over what is done. Good restorers normally have a waiting list, and the charges can be quite high depending on the work that is deemed to be necessary. Restoring a dial takes time and skill, so an advertisement which promises a quick job means a firm to be avoided.

In the absence of any written guidelines on conservation and restoration of painted dials, the following sections may be of use for the serious student, museum conservator or collector.

EQUIPMENT AND MATERIALS USED

The equipment needed by a dial restorer is not difficult to get hold of. The largest item is the dial restorer's table or stand (**Illustration No. 117**). This cannot be bought and must be made. It should be sturdy but moveable so as to adapt to differing light conditions. The revolving centre turntable should be quite thick, being made of several sections of plywood glued together. Various holes will have to be cut out in order to accommodate different positioning of dial feet, posts, irremovable springs, etc. Besides the normal travelling bridge with a stainless steel straight edge illustrated here, there should be a perspex or transparent thick plastic bridge which can be of great use when doing repeat corners.

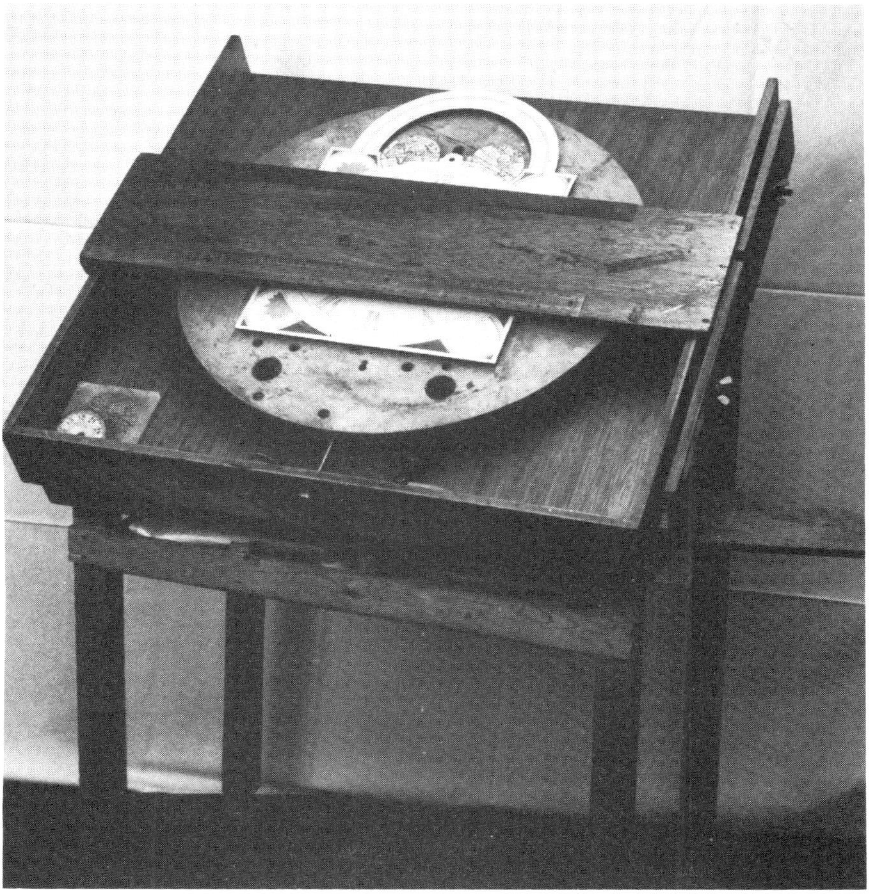

117 The working stand and revolving turntable with travelling bridge. See page 183

A frame stand (**Illustration No. 118**), which also is not available commercially, is of great assistance when marking out circles on the dial in conjunction with an adjustable centre hole locater. The frame stand must have non-slip feet.

The most difficult item to procure is the ultra violet lamp, which is absolutely essential. Over the years, some of the carbon from the black

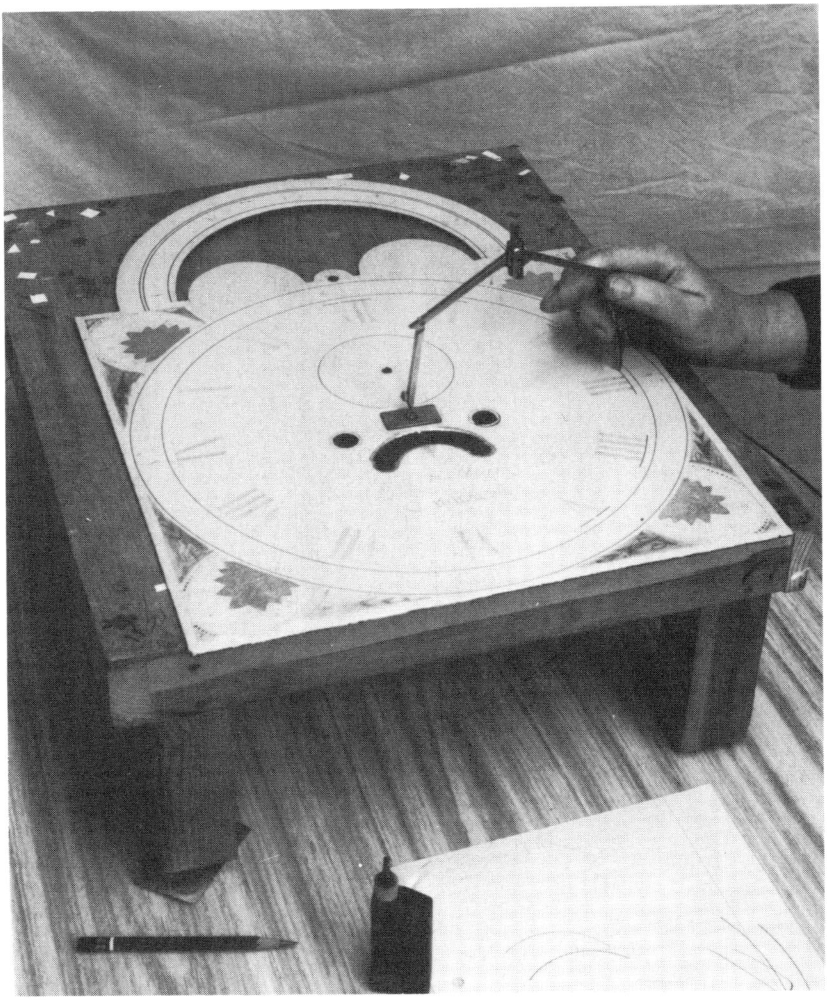

118 The use of the frame stand with centre hole locater used to mark out circles. See page 184

or graphics work on a dial will have sunk into the base paint. Consequently, although the detail may have been rubbed or washed off so as to be no longer visible by daylight, the carbon will fluoresce when illuminated by ultra violet light in a darkroom, so missing detail can be restored. The ultra violet bar lamp is not nearly so useful, being merely a fluorescent lamp coated with a UV emitting phosphor. It is necessary to use a Woods glass envelope which absorbs the visible spectrum and produces a powerful UV illumination. It is not advisable to move this lamp about a lot, as the element can be damaged. **Illustration No. 119** shows the set up. (The ballaster fitted with rubber feet forms the base of

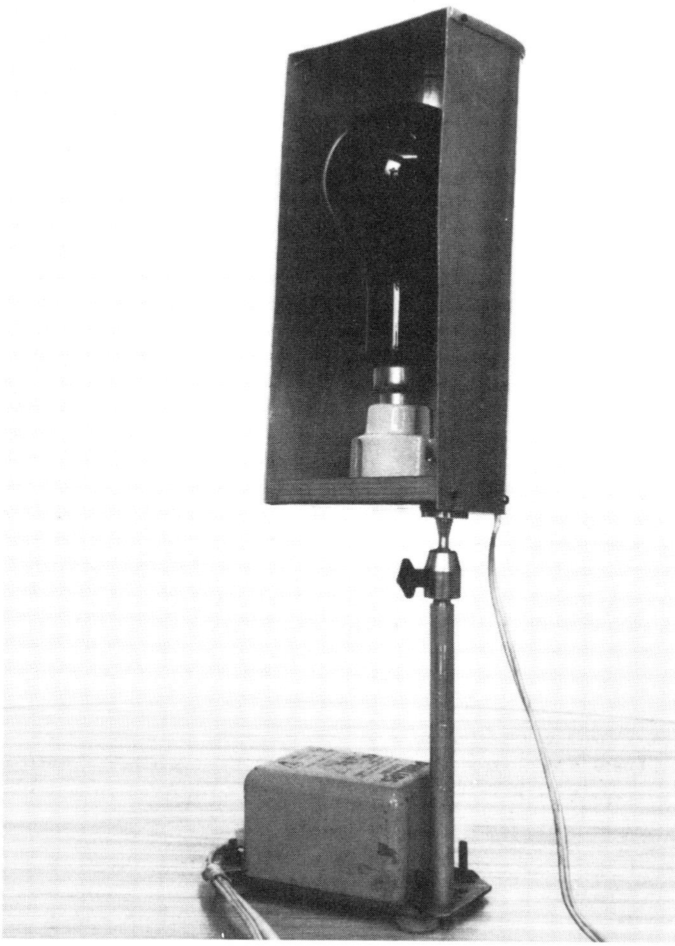

119 Ultra violet lamp complete with shade. See page 185

the lamp. This is a MBW-L5125BX. The lamp is a MBW U 125w with a three pin bayonet connection cap.) It should be noted that the aluminium reflector shields the operator from the direct UV light, which can damage the retina.

There is always some UV light reflected from the surface of the dial, and if it is being used for long periods it is advisable to wear glasses which will absorb some of this UV. If one is to be very particular about this, use glasses with clear glass which has a high cerium content, i.e. a UV absorbing glass. The darkroom should not be painted white nor should the operator wear white clothing because of the very strong dazzle effect. The UV lamp is a wonderful asset, but it does have its limitations. For instance, the depth of penetration of the UV is not great, and a really dirty dial can defeat it, until some of the dirt is removed. Also, some paints or inks do not fluoresce well at all. Large areas where extensive previous restoration of base paint has been done will show up at once, as will staining caused by oil absorption, usually around the dial centre.

The usual array of pliers, light hammers and fine to coarse small files are necessary, as is a small vice with soft jaws. Two electrical gadgets are an asset, but not absolutely vital. A Burgess Powerline heavy duty engraver, using the engraving point, will quickly dig out crumbling patches of paint and rust, while the rivetting head will redo any loose dial feet before the real work is begun. It cannot be said too often that it is necessary to closely examine all dial feet, posts, etc., especially in the case of long feet which often have worked loose. A mini 12-volt drill, using a fine grinding point, will smooth off the rough surface caused by the engraver, and it can also be used to drill out broken clock dial pins that will not come out by any other method.

The best set of drawing instruments available is a very good investment, as they can literally last more than a lifetime. The author is still using her father's instruments *c.* 1904. Rotring ink has proved to be more satisfactory than India ink, as it seems to be more permanent. The smallest crow quill or mapping pen should be used for the maps only. Brushes are very much a personal preference, but the Windsor and Newton sable brushes series 3A sizes 2 to 00 seem ideal. These are generally sold as watercolour brushes, but there is no reason why they cannot be used with oil paint. The size 0 is the best for fine work, but a new brush cannot be used for this right away as it has to be 'trained' by doing coarser work such as filling in of numerals etc. until it has lost quite a few hairs and the tip develops a definite bend in one direction. Some square ended sable brushes are useful for larger areas and fan brushes are used for the soft blending in of colour on the moon's face. Synthetic brushes are not satisfactory. A two inch good quality bristle brush is necessary for putting on the special varnish. Japan gold size is the adhesive used for both gold and silver leaf.

Solvents used are: industrial methylated spirits (IMS), surgical spirit, acetone, zylene, turpentine, ammonium oleate, white spirit and trichloroethane. Abrasives used are pumice powder, 800 grade silicon carbide, and very rarely commercial brass polish 'Brasso' which consists of an abrasive (chalk and china clay), white spirit, ammonia and fatty acid. Various grades of carborundum paper and steel wool are used for removing heavy corrosion from the back of moon and date dials. Powdered orris root is very good for absorbing grease left in tiny cracks on a dial surface. Hydrogen peroxide is a good bleach if used with care, and oxalic acid removes some rust marks.

Scalpel blades nos. 22 and 23 with rounded blades and no. 26, which is flat, are used without a holder for delicate removal of some types of overpainting and a Stanley slim knife is used for tougher work. The smallest size of pegwood also comes in useful when making up a swab, which should be made of best quality pure absorbent cotton wool covered with a square of clean white linen from an old sheet or pillowcase. Linen damask such as found on tablecloths etc. will not do. The finest linen, which is usually from old handkerchiefs, is for making 'fads' used with varnish, which will be explained later. White flannelette or Canton flannel is used with the silica carbide powder or Brasso.

The filler that seems the most satisfactory is a proprietary one, 'Elastic Plastic Padding', used in motorcar body repairs. Some practice in using this product is essential. It can be applied with a spatula for large areas or a small knife tip for the lesser holes. The special varnish used over the whole dial in order to seal it is Paraloid B67 dissolved in white spirit, or alternatively Rowney or Windsor and Newton Retouching varnish. Artists' gloss varnish can be used for coating decorative work when it is dry. For undercoating the filler only, and in small unfillable spots, Humbrol Enamel Matt is a quick drying paint providing a good key for the artists' quality oil paint which is the only other paint used. A list of useful colours in artists' oil paint include:

Flake White	Crimson Lake
Ivory Black (best for decorative work)	Scarlet Lake
Bone Black	Bright Red
Lamp Black (always used for graphics)	Viridian Green
Raw Sienna	Cerulean Blue
Burnt Sienna	Prussian Blue
Raw Umber	Ultramarine
Burnt Umber	Cobalt or The'nards Blue
Cadmium Lemon	Yellow or Gold Ochre

Thinners are real turpentine, linseed oil and gloss varnish.

Once the brass surface on the back of a moon and date dial is made as clean as possible, it is advisable to coat it with a suitable lacquer at once.

Ercalene has proved to be very good.

Siliconized vegetable parchment or silicon release paper is used to prevent the centre hole locater from sticking to the varnished dial, and white acid free tissue paper is used over a soft pad when the dial is completely dry in order to attach the false plate. A martindale face mask is useful when removing heavy corrosion on the back of a moon dial as the resultant powder can be quite irritating to the nose and chest.

TYPES OF DIAL DAMAGE (and how to recognize them)

INHERENT FAULTS AND PROBLEMS

In dealing with these, it is best to work from the bottom outwards. The wrought iron dial plate was usually in the earlier periods primed with a coat of iron oxide paint, which is ideal as it forms a good base for subsequent layers. When it has not been used, trouble is often encountered. Some very early dials appear to have employed a partly bituminous type of primer which never seemed to completely dry and, being plastic, the top layers of paint cracked open. This type of primer shows up brown black in the open cracks, and it remains soft. It has been used on some Manchester dials. Another type of primer found on later dials was a shiny black or red brown one, usually put on rather thinly. The shiny surface provided no key at all for the subsequent layers and frequent delamination was the result. Sometimes this is in the plain areas only, such as the chapter ring, leaving the parts with the artwork relatively sound, which could lead one to believe that delamination is more liable to occur where there was no protective artwork and varnish which covered the vulnerable paint. Incipient trouble of this sort can be seen in a raking light where evidence of cupping or distorted lifting of the paint layers is more noticeable.

Two different metals, such as iron and brass, have different rates of expansion and contraction. The rivetted-in brass feet often caused trouble early on in the wrinkling of paint just at that place. Sometimes the paint in that area is perfectly sound but other times it is not, and may have flaked off, though frequently this sort of damage is caused by the dial being dropped.

As regards moon and date dials, when they were made of brass, as about 80% were, no special undercoat was used. Date dials can have flaking paint but the moon dials' paint seem to be more vulnerable, and large areas of paint loss can be quite common. This can be noted in the painted moon dials on brass dialled clocks. The more highly finished and shinier the brass, the worse the paint adhesion. Another bad problem comes up when the back and front of the moon dial have been thoroughly cleaned and the dreaded stress cracking of the brass is noticed. Of course, really bad stress cracking, which in extreme cases can reduce a moon dial to a pile of fragments, will have been seen at once. It can, however, be almost invisible in the form of hair line cracks in light areas such as the sky, or it can be very severe indeed so that a replacement moon must be made. The cause of this cracking is inherent in the brass itself. The composition of brass varied greatly in the early period before 1810 or so, when most of the stress cracking is found. The continual hammering of the metal to flatten it produced a work hardened surface which, if not annealed, could start to crack up. Often stress cracked moon and date dials have shed teeth. Disintegrating moon dials should be measured, photographed and traced at once before they deteriorate any further.

Base paint also varied considerably in its composition. Occasionally a few Wilson dials show a type of spotting caused by insufficient mixing of the ingredients and the result is small pinkish round areas which tend to crumble when pressure is applied. This is a defect about which nothing can be done, as the clay filler has not been incorporated into the paint. There is a vast difference between the modern finely ground paint and the coarser mix used on these dials. Close examination of the old base paint will show a number of specks in it of varying sizes.

Corner and break arch decoration sometimes used gold or silver leaf as a background, with or without coloured translucent paint applied on the top. There are, unfortunately, quite a few instances of the whole area crumbling away, noticeably in Scottish-made dials. If examined closely, the leafed area has a network of very fine cracks all over it. The paint appears unstable and this problem has several causes. The Japan gold size or other adhesive used to stick down the leaf was either too thick or had not dried enough before the leaf was applied. Then a very heavy coat of varnish, often coloured, was put on, and in the case of Scottish dials, thick black detail paint went on top of that. The result of all this was the fracturing of the layers from the bottom upwards.

Heavy varnish causes a lot of problems in itself. Old type varnish yellowed easily, and heavy varnishing can literally pull the paint off. Where the varnish has gone over the edges of the artwork on to the base paint it can show up as a yellow halo, which has given rise to the fiction that the artwork was put on by 'transfers'. Some paint is inherently unstable anyway, and parts of mixed colours change and fade at different rates, giving an effect after 100 years or so that is quite different from what the original artist intended. Other colours crumble and flake off. A definite change in colour is often seen when the grommets are removed if the decoration extended that far.

It is quite usual for a painted dial to have cracks all over it. Sometimes the cracks are few, and in others it almost resembles crackle glazing.

Providing the dial is sound these cracks are normal, though a very few dials have no cracks at all. Judicious prodding with a rounded end of pegwood should produce no noise at all. If an ominous cracking is heard, it means the adhesion of the paint is poor. There is a type of defect that was caused by either poor preparation of the dial plate itself or poor dial storage, i.e. damp. This problem shows itself as a bubbling up or rust blister which can happen anywhere on the dial, though it seems to be more common on or near the bottom edge where the dial could have rested on a damp surface such as a stone floor. Some of these blisters look as if they developed when the dial was relatively new, as the paint film has stretched and not broken up. Other times the blister has broken, inevitably exposing rust underneath.

ACCIDENTAL DAMAGE

Accidental damage is that which has been caused by something untoward, and not done on purpose. One of the most common accidents that happen to clocks, especially longcase ones, is the falling over of the whole clock, or the dropping of dial and movement. Usually there is damage to the dial feet which can be loosened, and in some cases the whole foot has been wrenched off. This is quite different from the sometimes quite severe bending of long feet done by the original clock maker in order to make the dial fit the movement. Also, it is quite common to see the main dial plate bent or a corner twisted. Moon and date dials sometimes show evidence of being dropped onto a hard surface, because teeth are bent.

Water, such as from a burst pipe, is also very detrimental. Both ink and paint are to a degree water soluble, but worst of all is the propensity of the whole dial to absorb water if the exposure is prolonged. This can lead to expansion of the paint film, and then serious cracking as it dries out. Water damage usually leaves long streaks of dissolved paint and ink down the dial, though some of this can be found to have been altered by someone's efforts with a cloth to mop the poor thing up.

There are atmospheres which damage dials. Smoke and soot leave a dark brown deposit which can stain, and the oily haze created by continual frying at too hot a temperature makes an even greasy film all over the dial. This is not improved by the sporadic careless swipe at the dial face by a duster or damp cloth. Much more insidious is the action of salt sea spray which tends to make dials corrode. This can lead to the rust damage which was mentioned in the previous section on inherent faults. Rust is easy to see, as it lifts the paint film and often leaves the bare metal exposed.

Very often clocks were not moved when redecoration was done, and somehow paint splatters, usually limewash or 'distemper', got on to the dial. These tiny spots have a chalky appearance and usually come off very easily. The ubiquitous house fly also leaves its marks, small round spots that are hard to remove. These spots can actually eat into the dial surface. When a clock was moved from one house to another, any automata were likely to get lost, though previous clock repairers quite commonly removed these if they 'gave trouble'.

There are times when a clock is so situated that strong sunlight falls on the dial, which will eventually fade some of the more impermanent colours. Other times a clock is right next to a heat source such as a radiator and the convection currents of air carry an enormous amount of dust on to the clock dial which eventually gets 'dusted', leaving light random streaks. Then there is the careless owner or custodian who, when winding the clock or moving the hands around, scratches the dial with the fingernails, which carries the black pigment from the numerals on to the base paint. These scratches are concentric and usually strongest at the bottom of the dial. There is another type of scratching caused by the hands being too close to the dial or if the hands are pushed into the dial when moving them on. Key damage often occurs around the winding holes, and a poorly sighted or possibly drunken wielder of the winding key can cause a lot of damage.

Two world wars and other sorts of explosions lead to bomb damage where the hood glass is shattered, leaving minute fragments of glass embedded in the dial. This sort of thing is easy to spot, as the pieces of glass reflect the light, giving a glittering appearance. The injudicious use of a football in the house can also lead to fragments of glass in the dial, but in these cases the pieces are usually larger and, having done their damage, drop out. Improper storage can lead to other problems, such as bird droppings or contamination with strong acids or alkali which can eat into the dial surface. There is a modern type of damage which is hard to detect, but the use of a glass cleaning spray on the hood glass can affect a dial if used continuously, because it leaves a pale powdery deposit. Some of these sprays can contain ammonia, which is very bad for a dial. More insidious is the greasy vapour left by spray polishes on the hood which eventually settles on the dial, because clocks are anything but airtight. Occasionally the inner mask of the hood is polished with wax or furniture cream, and some of this inevitably gets on to the dial, leaving a mark which turns brown in time.

Before the days of electricity, clock makers worked, if it was dark, by candlelight. Blobs of spilt candle grease are quite common on the back of the dial, but sometimes the candle upset and the varnish on the front caught fire. This caused a localized bubbling up of the surface. Much more serious damage could be caused by a misdirected blowtorch or blowlamp flame, presumably used by a solder artist. Worst of all accidental damage is the fire where the whole clock was burnt. The dial is then covered with blisters of burnt paint and sooty streaks. In many cases the red iron oxide priming coat can be seen as this paint seems to resist heat better. The efforts to put out the fire usually mean water damage as well.

DELIBERATE DAMAGE

Deliberate damage is that which is done on purpose, though some of it may have been undertaken in the mistaken idea that the dial needed improving or modernizing. Scratches on dials are quite common but sometimes one sees the deliberate scratching of the decorative elements where it looks as if a small child had got at the dial with a large nail. The same applies to crayon and felt tipped pen marks which nearly always are on the bottom part of the dial as if the child could reach no further. Then there are auctioneers' marks such as 'Lot No. 3' done in large letters, or the sticky label sometimes denoting who was to have the clock upon the decease of the owner.

Oil of any sort is a dreadful enemy of dials. If it is thick enough and left for a long time it can actually destroy the base paint. It is most commonly seen around the centre hole and the winding holes, but it can be anywhere, sometimes running down the dial in long streaks. Date dials can be liberally soaked in oil, leading to the complete disintegration of the paint. The modern spray can of WD40 delivers a nice film of oil all over everything. The use of furniture polish on a dial can be classed as oiling. Brass grommets and even hands were cleaned with brass polish which eventually could wear a dial right down to the red oxide primer. This sort of damage by the over zealous housewife is usually located around the winding holes. The hand polishing was done further away from the dial because the ends of the hands had to be supported.

The urge to scrub something is obviously in many cases impossible to resist. Dials have come in for restoration where the merest ghosts of their former selves are visible. After the scrubbing with soap, detergents or abrasives comes the 'putting the numbers back'. That marvellous modern invention 'Letraset' sometimes is used, but more commonly some form of paint or ink is employed with varying degrees of expertise. The non-professional hand is easy to detect with its over heavy lines and hesitant Arabic numerals. Some of the scrubbing operations were seemingly done to remove one clock maker's name and substitute another. If this was done a long time ago, it can be hard to detect, but usually the UV lamp shows it up. Not only were the graphics redone but often the decorative elements received improvements as well. Far more difficult to deal with is the dial where the whole surface was overpainted. The dial usually looks anything but flat when this was done. Sometimes a heavy coat of varnish was put all over the dial in order to give it a nice shine.

There was a time when it was considered easier or better to supply a new or different movement instead of repairing the old one. New winding holes had to be drilled and the original ones filled in, or else long feet could be twisted or bent. In the case of bent or long feet, as mentioned before, sometimes the original clock maker bent the feet in order to fit his movement, but where a new movement has been supplied either it does not look of the same age as the dial or other things do not match up, such as date or seconds holes. Slightly oval winding and other holes were quite common in dials with original movements. It could be considered that all this is unimportant, but obliterated seconds rings and drastically altered winding holes mean something is not quite as it should be.

Unfortunately, some dials do get shot at, and at least one dial encountered had a hole right through it, thoughtfully mended with an adhesive plaster. Another boy obviously knew not to shoot at foxes, for only the rabbits were peppered with holes and the foxes on the dial remain untouched. Where a moon dial has been shot at, the brass, which is much softer than wrought iron, can take quite a beating, and the moon dial ends up very pockmarked, often to such a degree that it will not revolve until the 'dents' are flattened out.

There are bound to be other manmade afflictions of painted dials, but this survey hopefully covers most of the more commonly seen disasters. Until the painted dial receives the respect it is due, the horror stories will, unfortunately, continue.

HOW TO REMEDY DIAL PROBLEMS

INHERENT FAULTS

Having discovered what is wrong with a painted dial, the next step is to remedy the ailment. Dials do not fit into neat classifications as they often have several things wrong with them due to different causes. Thus the already delaminated dial left on a rubbish tip can have been shot at as well, or the dial residing in a cow byre has had birds perching on it. Sadly, not everything can be remedied, and so a repaint may be, as a very last resort, deemed necessary. Repaints are really a separate subject which will be dealt with later. Ideally, in the case of a repaint, a whole new dial should be made but often the cost of this is prohibitive.

Paint Incompatibility and Delamination – A Step-by-Step Process

In many cases it is not possible to determine just what has caused the damage. Something like delamination or rust can have many contributing factors. A really bad case of paint incompatibility

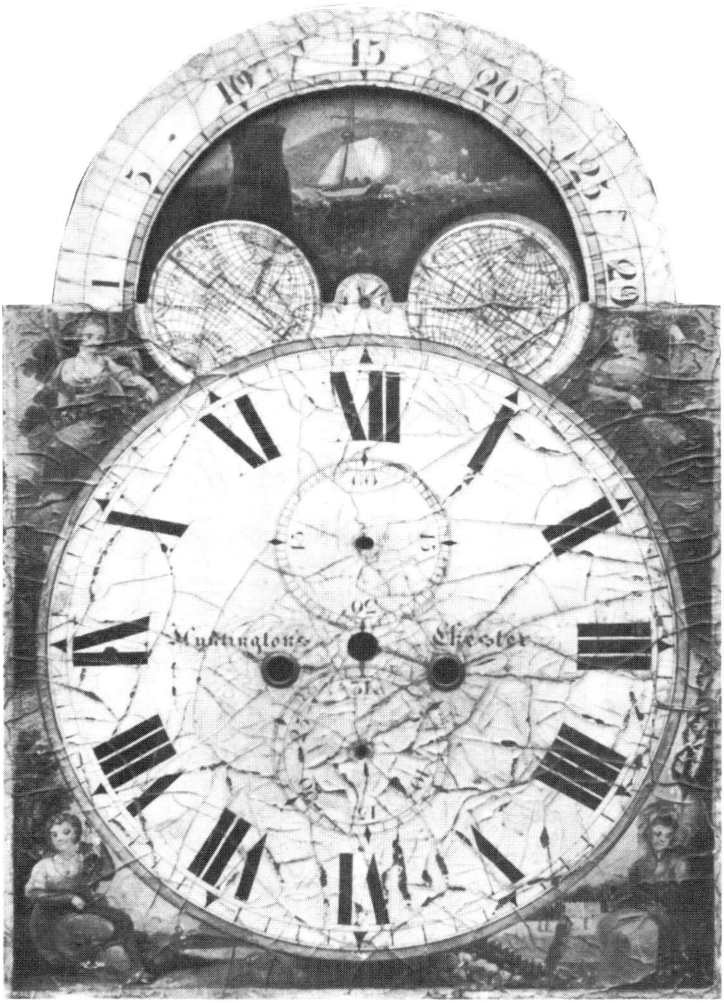

120 Bad paint incompatibility has almost destroyed this Huntington dial *c.* 1840–50. See page 194

(**Illustration No. 120**) is seen in this Huntington dial from Chester which, though cherished in the family since it was bought, ended up looking really dreadful. This dial (*c.* 1840–50) is one of the four Huntington dials known to the author, and the dial maker is unknown. The base paint, not iron oxide, appears to have shrunk, causing the buckling, detaching and crumbling of the top layers (**Illustration No. 121**). Nothing could be done with this dial except a complete strip and repaint (**Illustration No. 122**). Luckily, not all of the Huntington dials suffered this problem, as another dial of this type is illustrated

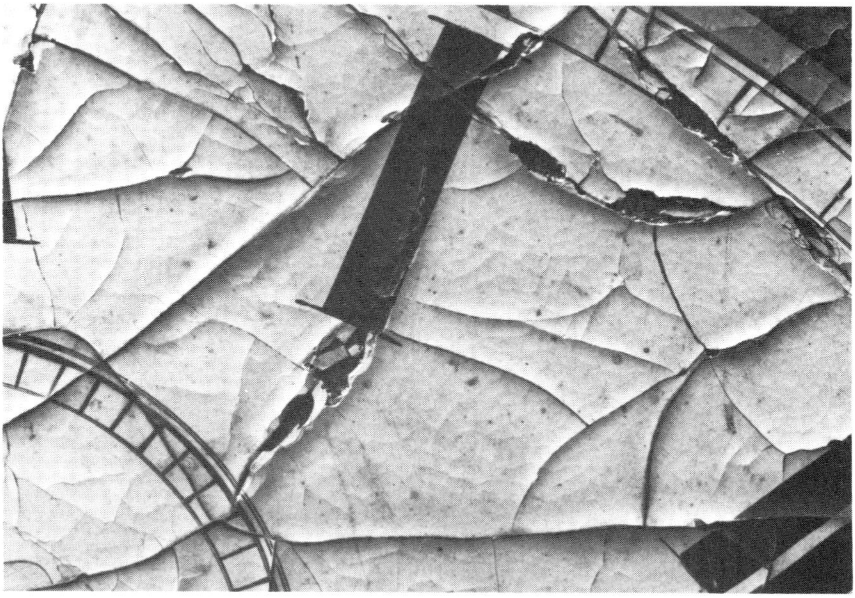

121 This close-up shot of the Huntington dial shows the extent of the cracks. See page 195

(**Illustration No. 61**) in the earlier section concerning unknown dial makers.

Even Osborne and Wilson made mistakes, and this (**Illustration No. 123**) is the result of their using the wrong primer in 1772. The paint has not only delaminated in many areas where the black primer can be seen, but other places are very badly cupped. On top of all this someone a long time ago washed the dial and put the numbers back. It was also oiled with a dripping oil can. This dial was a real challenge as it was an important early one in a dated case (1772). Further details concerning this dial plus an illustration of the finished restoration will be found on p. 28–9, **Illustration No 5**. The restoration of this dial will be gone into in some depth as an example of really bad deterioration.

First of all came the photography in colour, and then a careful tracing of the whole dial as it was, repainted numerals and all. Next was an examination under UV light, and the putting in of the lost name using a 6B pencil with great care because of the delicate state of the base paint. Another tracing was done after that. Then the real work could begin, but notes were taken of labels, scratched repairers' marks and anything else on the back of the dial. Surface dust was removed with a very soft brush such as camel hair. Next, possible solvents for the overpainted numerals were tested on the back of the dial where the ground paint had slopped over. The most satisfactory solvents for much overpainting are surgical

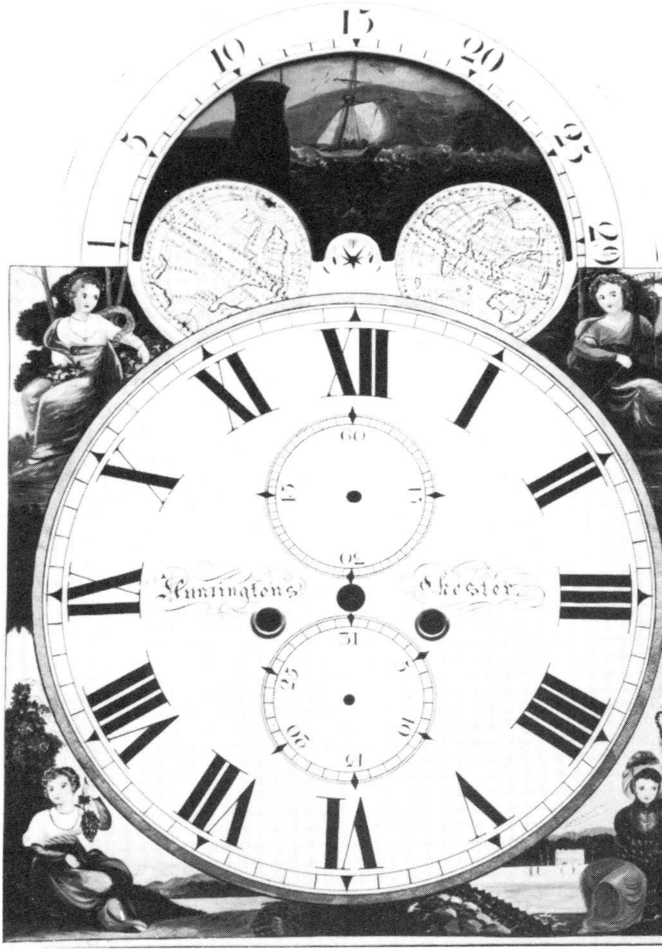

122 The Huntington dial has been completely stripped and repainted. See pages 194 and 195

spirit or industrial methylated spirit but even these can attack some base paint. If the base paint comes off on to the clean linen covered swab, leaving a sticky patch, one has to think again. Base or ground paint, even with Wilson dials, was anything but a standard mixture. Luckily, these solvents mentioned do not attack most early dials. The next step was to de-grease the dial with 'Gendene' (Trichloroethane), which can also in some cases help to soften overpainting. Then came the removal of the overpainted numerals with a great number of already prepared swabs. It is most important to remember that this work must be done consecutively as one is removing old black paint which can settle quickly

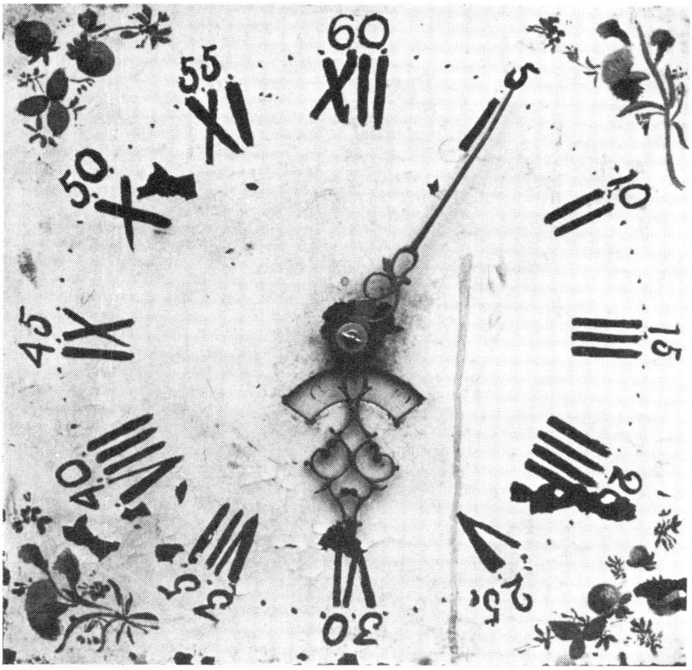

123 An Osborne and Wilson dial of 1772 with just about everything wrong with it. There is an extreme case of de-lamination in some areas as well as oil damage and clumsily repainted numerals. The restored dial can be seen in Illustration No. 5. See page 195

into the cracks and harden. About one or two inches square is the limit of the area to be cleaned at one go, so a whole day is usually allocated to this process. Sometimes a scalpel blade of the rounded type can be used with the greatest care in order to remove some of the thickest of the black paint.

After the very tedious removal of the overpainting, the whole dial was cleaned with the same solvent, as the rest of the dial was pretty dirty. Any remaining pencil marks were removed by the solvent. Great care has to be taken with the artwork as some colours can come off very easily, red being one of them. Another examination by UV light followed, when all of the correct numerals as well as the name were very lightly pencilled in using the 6B pencil. Names and towns are notoriously prone to disappear, so that is why separate early tracing of these was necessary. Another very detailed tracing was then made, as the cleaning and de-greasing had revealed not only the original graphics but details of the missing artwork. A light going over with the solvent in order to remove pencil marks came next, followed by the use of fine pumice powder to remove some of the residual staining of the overpainting and oil damage. Solvents soften the surface of a dial, so it was necessary to allow the dial to harden for several hours before using the abrasive. After this came the careful washing of the dial with green soap and warm water followed by thorough rinsing and immediate drying with a soft cloth. Very sound dials can have the final rinsing done with quite hot water, increasing the temperature gradually and drying at once with a polishing motion. This often reveals a lot of hidden detail. A washed dial should always be left to dry overnight in a room with good air circulation.

The next day it was possible to see what needed filling, depending on the depth of the missing areas of base paint. It was decided not to attempt to do any 'plastic padding' filling on this dial as the base paint was quite thin and very fragile. The next step had to be the priming coat of a quick drying undercoating (Humbrol Matt) in a toning colour of all areas that showed the black primer. The dial was then left for a day before starting the next process. A quantity of artists' oil paint was mixed up to match the base paint, and brushed carefully all over the dial, giving as flat a surface as was possible. No thinners at all were used. Dials often are different colours in different places, so matching is essential. A useful colour base for early Wilson dials as well as Osborne ones consists of: flake white, cobalt blue and raw sienna. The dial was then left for roughly 12–24 hours in a room that was as dust free as possible. After this the paint layer, which in this case covered most of the dial, was gently smoothed flatter with the fingertips and special attention was given to the cupped areas. It is important not to leave 'fingerprints' in the paint. If the paint has already dried into brush marks, one is too late, and that paint layer must dry thoroughly before applying another thin film of paint which must be smoothed at the right time. It is very difficult to explain just when to catch the paint at the right moment, but this smoothing process seems to help the paint get right down into the cracks. This application of paint was done at least twice on the 1772 dial, with weeks of slow drying in between. It is inadvisable to use a heated drying cabinet or rack at this stage, as further cracking may result.

The dial did look very odd with paint all over it, but when the paint was really dry after several months, most of it was removed with spirit (IMS or surgical) on linen covered swabs, except where it covered missing paint areas. The cupped areas were found to have flattened out considerably. Further inpainting was necessary in the places where the original paint had flaked off, and the surface was made as flat as possible. It is most important that a dial is completely dry before the next step, and this drying process can take about three months, depending on weather and temperature. Once the dial was dry, the missing gold leaf could be put on, as the UV light had shown where it was. Should this dial have been embellished with gilded gesso, the gold leaf for this would have been put on at the same time. Replacement of missing gesso will be dealt with in the section on repainting or making a new dial. When the gold leaf was dry, a matter of several days, it was varnished with a clear gloss varnish in order to protect it. Another examination of the dial under UV light was done at this stage in order to make sure about the placing of the visible graphics. Although an accurate tracing was made, it is easier to indicate with a light pencilling where the graphics are to go directly on to the dial, and use the tracings to locate the graphics on any inpainted areas. Roman numerals on inpainted areas were pricked out using a scriber.

Varnishing was the next step, where the special light retouching varnish was used all over the dial, which formed a barrier between old work and new, except for gold leaf work and the reconstruction of the basepaint. The varnish was put on in a dust free room, with a good quality clean 2′′ bristle brush. It is inadvisable to varnish on a windy day, as inevitably the wind stirs up dust particles which then settle on the dial. The varnish was brushed out in two directions in order to form a thin layer, avoiding puddles of varnish in low areas. Sometimes a dial seems almost reluctant to take the varnish in which case a very fine linen covered swab or pad is used in different directions to try and persuade the varnish to stick. Much trouble can be caused by residual oil remaining in the cracks of the dial, and this oil can be almost impossible to remove. The varnished dial was left for the usual period of time, about three hours, for drying.

From now on, the dial had to be handled by the edges only, as the varnish remains plastic for a very long time. The dial was placed on the frame stand, and a piece of siliconized vegetable parchment was cut out to extend ⅛′′ beyond the edges of the adjustable centre hole locater. Finding the original centre used is never easy, because often it is anywhere but the centre of the hole for the hands. The locater had to be moved around until the compass lines fitted. Some of the old instruments used originally were worn or inaccurate so that often a slight compromise may be needed. The nearer to the centre of the dial the concentric circles are, the easier it is to find the true centre. The lines of the tops and bottoms of the Roman numerals could vary because the compass altered, so the originals must be followed. If there were circles on the outside of the chapter ring with gold leaf in them, as found on later dials, these circles could have used a different centre. Only the lightest of touches should be used with the compass pen, and no mistakes can be rectified. If this dial had a seconds and/or date ring, it also would have been put on at this time, using the locater with a fresh piece of paper to prevent sticking. Moon arch circles are more difficult. If the moon dial post can be unscrewed a short piece of pegwood should be put into the hole, and by trial and error the centre that was originally used will be found. Should the post be immovable it is necessary to cut out a small square piece of thick elastic sticky tape and place it on the rivetted end of the moon post. Usually two pieces of tape are needed if the post is very domed. This tape must be removed immediately once the circles are done.

Next, the dial was removed to the large turntable stand for the marking out of the rest of the Roman numerals, using the sliding bridge and an aligning pen. Looking at the illustration of the stand **(No. 117)** one can see a number of holes have been cut in the heavy plywood turntable, mostly to accommodate various dial feet and posts, but there can be awkward springs or automata posts at the back of the dial, and if holes had to be cut for all these, the turntable would soon resemble a colander. The answer is to pad the back of the dial with small blocks of wood attached by sticky tape which raises the dial clear of the obstructions, and an extra long screw in the centre is used to keep the dial in position. The travelling bridge may also have to be padded, in order to clear the dial. Dished dials which were used much later than the flat ones mean using the turntable in its lower position by removing the brass spacer between the back of the turntable and the flat base of the stand. Dished dials are inevitably more difficult to mark out.

If this dial had had maps, they would have been put on at this time before the varnish dried out and became brittle, but the maps really are a separate subject to be dealt with in another section. The Arabic numerals came next, using the series 3A Windsor and Newton sable brush size 0 with artists' oil colour lamp black. These Arabic numerals, together with the signature and any scrolling, are the most difficult graphics to do. The author remembers well the maxim of an old Chinese teacher of graphics who said 'Each letter or number must be a poem in itself', so there can be no hesitant strokes or flats on curves. The scrolling was done with a brush well advanced in its 'training' so that there were about half the original number of hairs.

In the case of this dial, some of the decorative work was missing right from the beginning, and the efforts to consolidate the base paint obscured more of it. Everything that had obviously been there was put back, matching colours exactly, and paying great attention to the type of brush strokes used. It can be difficult when working on several dials in succession to alter the type of painting done, but the original artist's style must be adhered to, even if it looks crude or the design seems unbalanced. When all the artwork was thoroughly dry, it was revarnished with artists' gloss varnish carefully following the design. The dial then had to dry out thoroughly for a month or so, depending on the weather. The graphics on some dials, especially where there is quite a lot of retouching varnish when a pad has not been used, exhibit a type of retraction of the black paint film which then looks rather cracked. The only cure for this is to wait until the dial is thoroughly dry, and then go over all the defective work with black oil colour mixed with a small amount of gloss varnish. It does not work if the varnish was mixed with the paint in the first place, as the same type of cracking will happen.

Date dials should be cleaned at the same time as the main dial, paying special attention to removing all the corrosion on the back, followed by a suitable lacquer. Date dials can require·filling and inpainting, especially where there is a lot of damage, and the same retouching varnish and paint is used as was done for the main dial. The finished dial and its date dial was photographed in both colour and monochrome before it was returned to the owner, but if the dial had originally had grommets they would have been put on after the dial was photographed. These grommets are often very fragile, and sometimes have to be completely replaced. Early flat Wilson type ones are easily made on the lathe, but the later rounder profile grommets can be bought and the outside edges turned down to suit the measurements of the originals. It goes without saying that the brass grommets must be cleaned and lacquered. Some dials have odd grommets for whatever reason, and the placing of these must be adhered to marking each one on the inside 'L+R'. When replacing the grommets, the dial must, of course, be supported by one person while another puts them in, using a shaped wooden peg held in the bench vice, and very carefully tapping and bending the prongs so that the grommets do not move about. If old grommets have to be replaced, the old ones, or what remains of them, should be returned to the owner. The restorer/conservator must put his or her name, address and date on the back of the dial, being careful not to obliterate any scratched markings.

Practically every dial owner is interested in its history, so the report compiled from the work notes should be as comprehensive as possible, giving details of what was done and why, together with advice on handling and future treatment. The author has found it better to use simple terms, for though museum staff will understand the use of 'verso' for back, other people may not. It is essential to keep the original work notes together with the number of the negative or slide, in a manner where they can be easily referred to. The author remembers one occasion when a hard pressed CID detective wanted instant details of a stolen clock over the telephone, and the only information given out was the clock maker's name, together with that of the rather distraught owner. The practice of keeping 'spares' of colour slides came in useful in this particular case, though as far as is known the clock was, unfortunately, not recovered. Sometimes an owner wishes to have a 'before' photograph, especially where the damage was as considerable as in the Webster dial. Indeed, looking at **Illustration No. 123** and comparing it with the finished restoration **No. 5**, there seems little similarity between the two, except for the artwork in the corners.

Non-Drying Primer

Another quite different problem is where the priming coat contained a proportion of non-drying paint such as bitumen, which, though adhering well, remained plastic and moved, thus badly cracking the base paint which also could have been overheated when drying. (**Illustration No. 124**). This Manchester-made dial *c.* 1785 was made for Samuel Collier of Eccles (1750–85+) and the photograph shows the dial in its original state when brought in. The large reticulated cracks can be seen in detail (**Illustration No. 125**) as well as the brown staining of the base paint near the cracks. The amazing thing about this dial was that in spite of all the cracking the paint layers remained firmly attached to the dial plate. Luckily, no one had attempted any treatment of this dial, and it

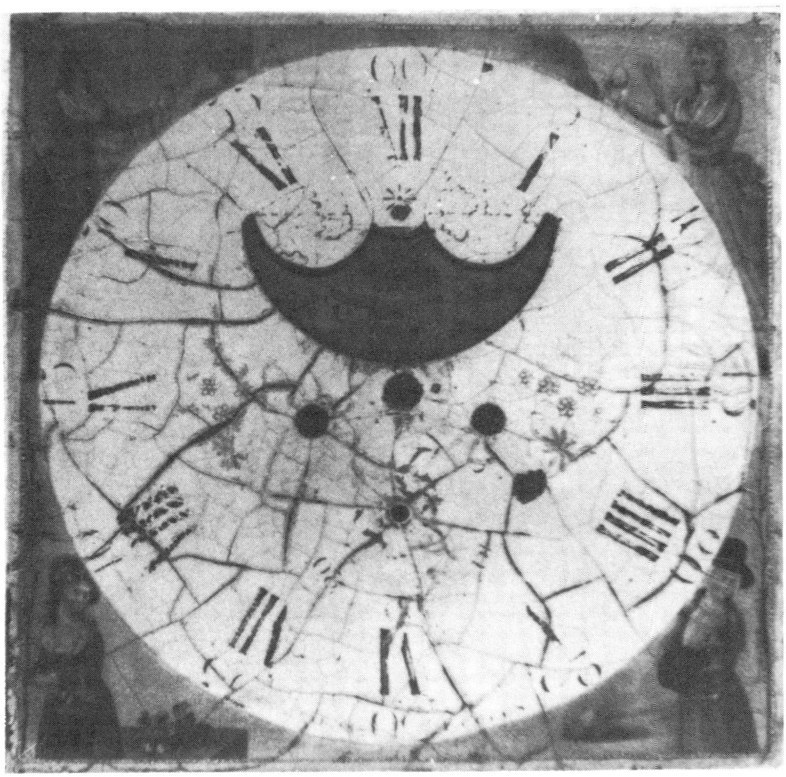

124 Unrestored Manchester-made dial used by Samuel Collier of Eccles *c.* 1785. The wrong primer containing a type of bitumen had been used and this caused severe cracking. See pages 202 and 203

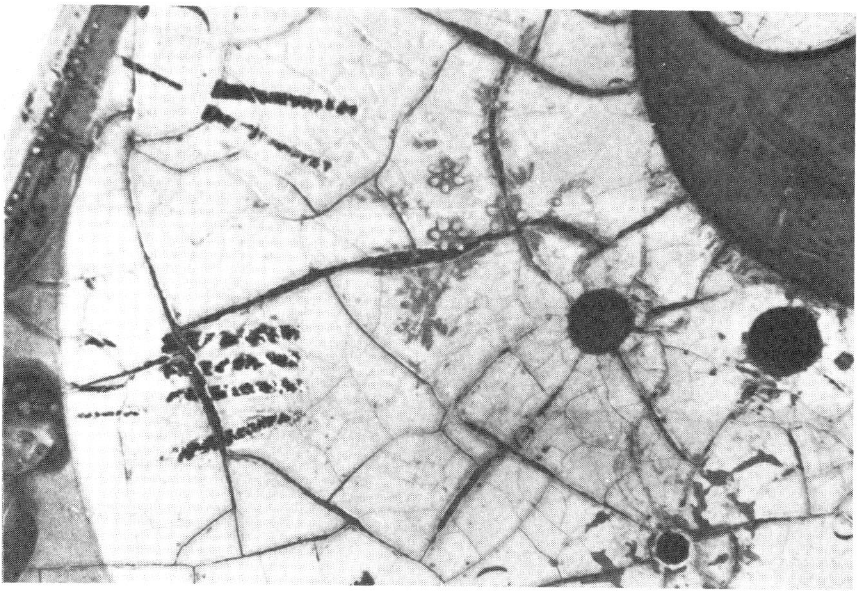

125 The detail of the Collier dial shows the depth of the cracking. See page 203

appeared to have at least a century of dust on it. The thorough cleaning of a dial like this obviously took some time, but a great deal of the dirt and all of the old varnish obscuring the corners was removed, though some of the residual staining remained (**Illustration No. 126**). It is quite easy to see the ghost images of all of the graphics, with the exception of the very small ones. Because of the variation in the height of the edges of the cracks, it was decided not to use a real filler. A toning oil paint, built up in layers, provided a satisfactory solution, though it took a long time for these layers to dry out. In a case of this sort, a sealing varnish had to be used wherever the priming coat was visible in order to prevent further staining. With this particular dial, the black priming paint was still slightly soft at the base of the open cracks. The final result of all this work was a very attractive dial (**Illustration No. 127**) where the four seasons ladies can be clearly seen against a pink background, and the strange Manchester-style maps are very evident. The use of a spray of gilded gesso flowers in the dial centre is rather unusual. Even in the photograph the alligator-like surface of the dial is quite noticeable, and some of the staining around the cracks is very visible. Cracks of any sort in a dial seem to worry the general public, and the author has been asked many times why she did not 'sand it all down flat and fill it'. Of course it has to be pointed out that one is not dealing with a front door, but an object of some considerable age. In any case, repairs and

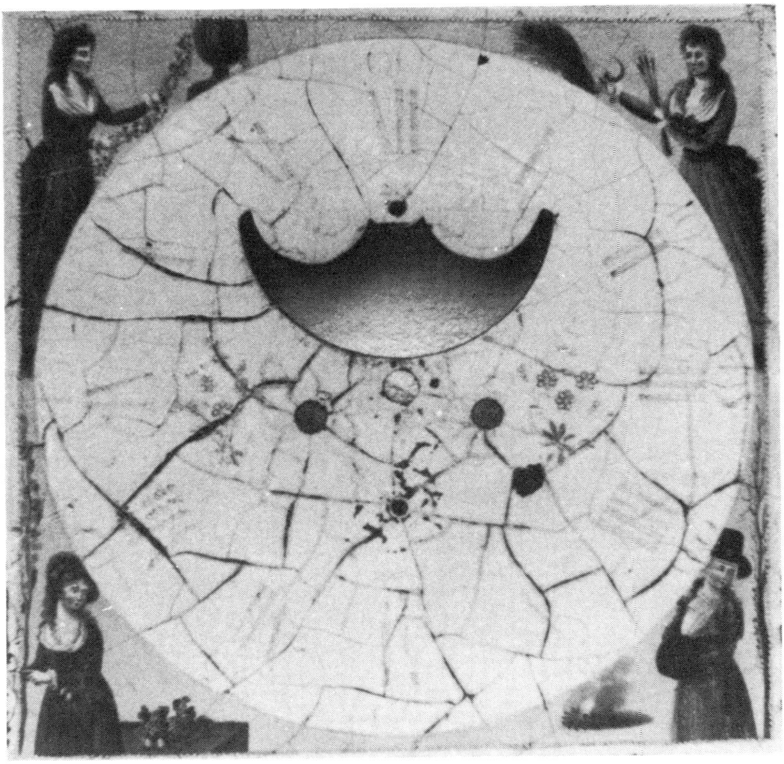

126 Now that the Collier dial has been cleaned ghost images of the numerals can faintly be seen. See page 203

inpainting should be kept to an absolute minimum, retaining as much of the original as is possible consistent with having the dial stable and not deteriorating any further.

Stress Cracking

The problem of gold and silver leaf which has either come off or become unstable will be considered later on in a dial that suffers from several other ailments. As mentioned before, stress cracking, which is often found on date and moon dials, can be very difficult to deal with. It cannot be said too often that the cleaning off of all the corrosion on the back of the brass dial is most important, and because it usually is such a very dirty job it is best to do it first. Next should come the removal of all the old varnish in the case of a moon dial. This can prove to be quite difficult, as it depends on the consistency, thickness and age of the varnish used, but its removal usually is quite dramatic when the delicate tones of the moon face are revealed and the bilious yellow look is gone. The moon dial (**Illustration No. 128**) from the Porter Wilson dial (**Illustration No. 44**) has been partially cleaned and the linen cloth on

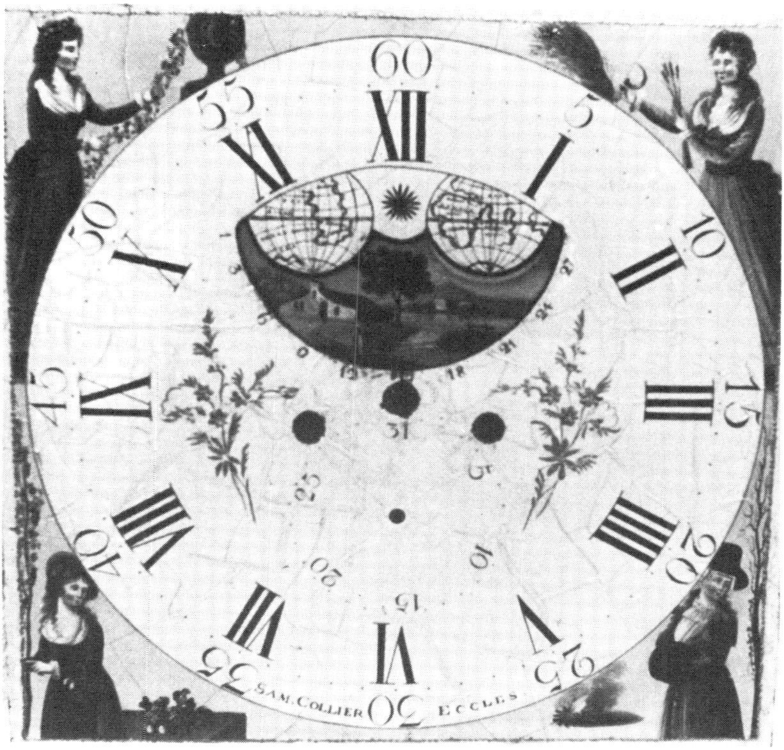

127 Restoration of the Collier dial now completed. The four seasons ladies are against a pale pink background and there are floral sprays of gilded gesso in the centre. The maps are the usual hand-drawn Manchester ones. See page 203

the left shows some of the varnish removed. Care must be taken on the black or very dark brown areas as the original paint was probably mixed with varnish, and because of this the dark areas can come off easily once a solvent is applied. Also, a very fine swab must be used around the edges of the moon's face in order to avoid dissolving the dark paint and transferring this to the delicate pink colouring. One easy way to find out if there is still varnish remaining on artwork is to touch it lightly with the fingers just after applying some solvent. If it feels sticky, old varnish is still present. Any hint of the colour, not yellow brown varnish, coming off, and the process must stop at once. The use of a scalpel blade on a moon dial is not advised because, due to the lack of a primer, the adhesion of paint to the metal of these dials is poor. There is usually a heavier build up of varnish around the teeth of the moon dial, and this can be impossible to remove completely. Naturally once the clean moon dial is dry and the paint has hardened up, a process which can take a day, new varnish must be applied. If there is no paint loss and no inpainting to be done the ordinary gloss varnish can be used, but where there is

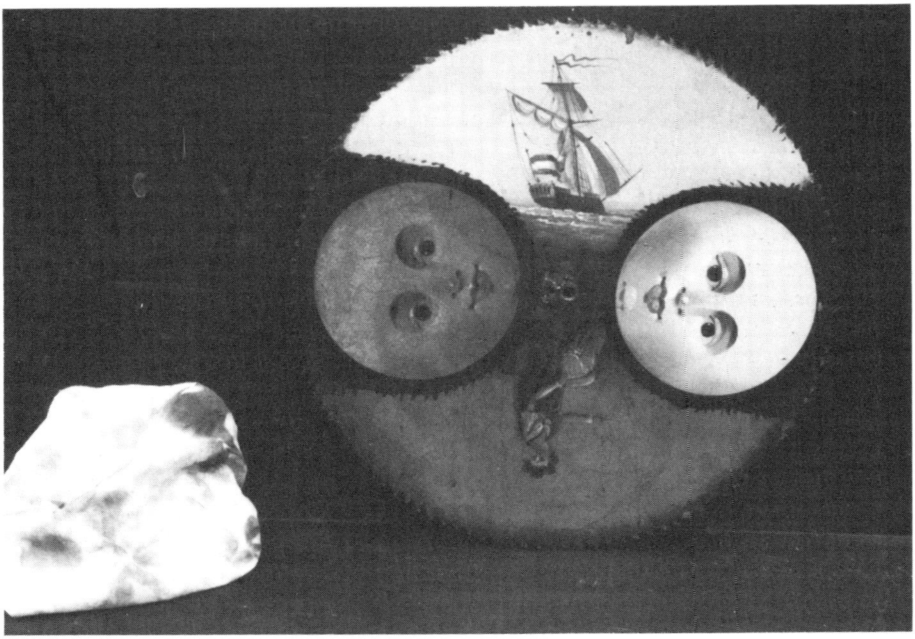

128 Partial cleaning of the Porter–Wilson moon dial showing discoloured varnish on the linen cloth at the left. The full dial can be seen in Illustration No. 44 on page 80. See pages 204 and 205

work to be done, the retouching varnish must be applied before doing anything more. The back of the moon dial should be cleaned down to the metal before the dial is varnished in the case of brass moon dials, and iron ones that have some rust on them may need some anti-rust treatment as, being much thinner than the main dial plate, rust could eat through the metal. It is also advisable to clean the old varnish, oil and dirt from the teeth of both moon and date dials, holding the dial vertical and using an old scalpel blade, but care must be taken not to dig into the metal or bend the teeth in any way. Some teeth may have bent previously, but if the moon or date dial functions, which should have been checked earlier, it is usually best to leave the teeth alone, as they can, under some circumstances, snap off completely. The application of a suitable lacquer to the back of a brass moon or date dial helps to inhibit corrosion.

ACCIDENTAL DAMAGE

Fire

Accidental damage is that which has been done to a dial without intent to alter or improve the dial. Fire is the biggest enemy of a painted dial. It can bubble up all the paint and melt the varnish so that it is almost impossible to see what was really there, which means a complete strip and repaint, or it can be limited to one or two areas. In the case of the really burnt up dial which has to be stripped, first comes the usual photography in colour and tracing, but after that the dial surface can usually be cleaned of smoke by using a solvent, and then the dial should be coated with quite a thick layer of retouching varnish. This usually reveals more details, so the dial must be rephotographed, and any details added to the tracing. Then comes the unfortunate stripping of the dial, after noting anything interesting on the back of the dial. The author has found 'Nitromors' water soluble stripper quite suitable, combined with an old brush and a scraper. The whole business is very messy and should be done out of doors using suitable gloves and protective clothing. Once the dial is free of paint, front and back, it should be well washed, and the final rinsing ought to be done with hot water. Very often the metal is found to be rough, so that it should be smoothed down using a sanding machine, though it is possible, though very tedious, to do this by hand. It is a good idea, if the metal is at all rusty, to give a treatment with a rust inhibitor, which usually is a form of phosphoric acid. Large holes and pits can be filled in using 'plastic padding', but it is essential to make sure all the foot rivets are secure. After this the procedure is the same as described later in Making a New Dial.

Partial fire damage (**Illustration No. 129**) can be bracketed in the same class as strong acid or alkali attack, or blow torch accidents. The paint bubbles up, and bare metal may be exposed in some areas. This particular Wilson dial suffered fire damage in the 1914–18 war, but was not restored until 1988. All the loose or blistered paint must be removed and the whole dial made as clean as possible, but inevitably there will be some discolouration (**Illustration No. 130**)). Strong sunlight is often a very good bleach, but it does take a very long time. Commercial bleaches must be avoided, such as 'Parazone', which contain chloride of lime. The holes must be filled and levelled, then coloured in as described in the delamination sections, pp. 197–9 & 223. Obviously in a case like this fire damage has left a lot of holes right down to the bare metal in some cases, so the amount of filling used is quite considerable. Once a flat and coloured-in surface is attained, any missing gesso should be replaced. The formula for this will be given in 'Making a New Dial', p. 242. Once everything is really dry, the gold leaf can be applied, and after using the retouching varnish the procedure is exactly the same as

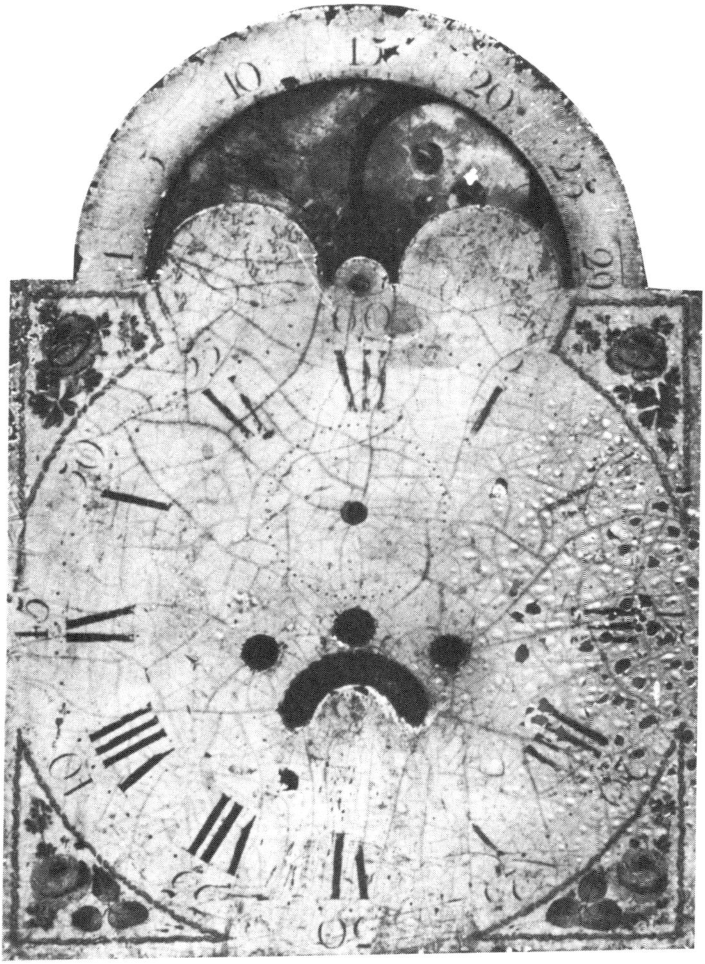

129 A Wilson dial *c.* 1785 that suffered fire damage during World War I. See page 207

that described in dealing with the delaminated dial. The results (**Illustration No.131**) show a Wilson dial *c.* 1785 made for the Chaplin brothers of Bury St Edmunds (1776–1810). Note the 'alternative' spelling of the name Chaplin. The dial will never look anything less than damaged, but it is now an acceptable item. A word of caution might be advisable here about limiting inpainting to damaged areas only so that as much as possible of the original surface is left. It is very difficult to colour match base paint, and the author has seen several instances in previously restored dials where large areas of perfectly sound paint have been painted over to cover just a few small damaged areas.

The moon dial of this Chaplin dial had not been damaged too much

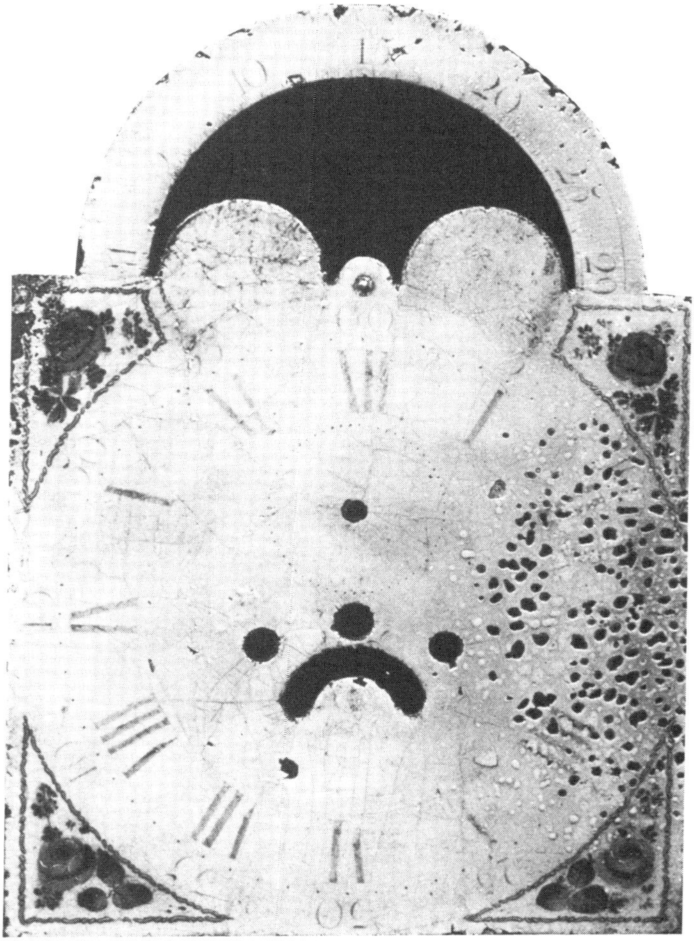

130 The Wilson dial cleaned down, showing bare metal in round holes on the right. See page 207

by the fire, but it unfortunately suffered badly from stress cracking, so much so that a great number of teeth had broken off. A skeletonized moon dial, following exactly the shape of the original teeth, was made and rivetted on. The very small rivets were not noticeable once they were painted over, and luckily the mechanism could be adjusted in order to accommodate the extra thickness of the moon dial. Needless to say all of this work on the moon dial had to be done before any restoration whatsoever was attempted, though photography and tracings were done in case the rivetting in such a fragile article caused the moon dial to disintegrate completely.

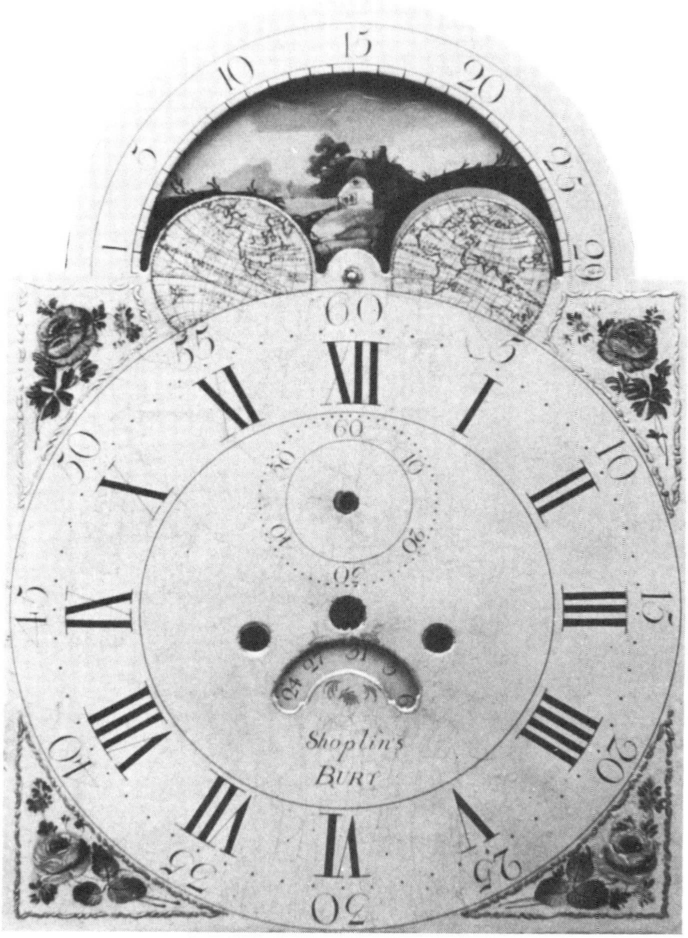

131 The restoration of Wilson dial is complete. It was made for the Chaplin or 'Shoplin' brothers of Bury St Edmunds. See page 208

Water

Water damage is usually streaky and can cause a type of bloom on any varnished areas which will look rather white. Usually the removal of the old varnish cures this problem. Also water can be absorbed to some extent by the paint film, and when this eventually dries out it may lead to cracking, though by the time the restorer sees the object it is not always possible to say that water has caused the cracking. Another probable type of water damage is where the base paint layer seems to have become porous and chalky. If this is left alone the surface usually deteriorates further, so in this case it is advisable to use inpainting on these areas, where the damage often occurs in long patches.

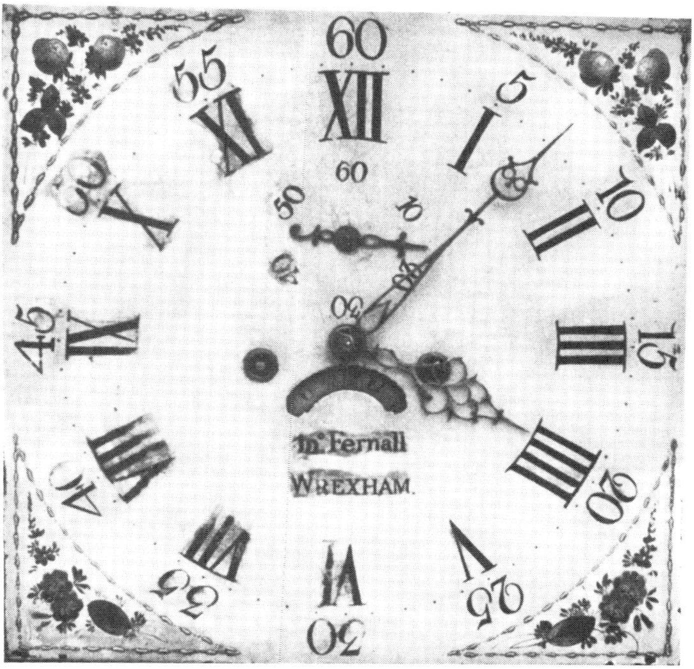

132 An Osborne dial made for John Fernall of Wrexham *c.* 1780 showing bad atmospheric pollution as well as the effects of partial washing followed by badly repainted graphics. See page 211

Atmospheric Pollution

Atmospheric pollution can change the colour of a dial completely (**Illustration No. 132**). This dial was made by Osborne for John Fernall of Wrexham (1770–89), and it dates from about 1780. The dial has suffered considerably from being in a very smoky atmosphere. It had been washed at least once and there had been a crude attempt at putting the numbers back. Luckily, it was a first class object to begin with, so it has restored very well (**Illustration No. 133**) and is a good example of typical Osborne pattern gilded gesso work framing the delicate fruit and flowers in the corners. Sometimes a dial had been very thoroughly washed a long time ago and then abandoned, though the original reason for the washing could have been water damage, as under UV light the dial revealed a lot of longitudinal streaks while the marks of washing were horizontal (**Illustration No. 134**). The dial was made by Walker and Finnemore (1801–11) for Thomas Bartley of Denbigh (1772–1811). The corner design was a Finnemore one as virtually the same corner appeared on a Finnemore dial made for Rees Williams of Cwmwysg (Brecon) (1737–1827) which has been shown partially restored in **Illustration Nos. 119** and **120**, and fully restored in **No. 118**. The

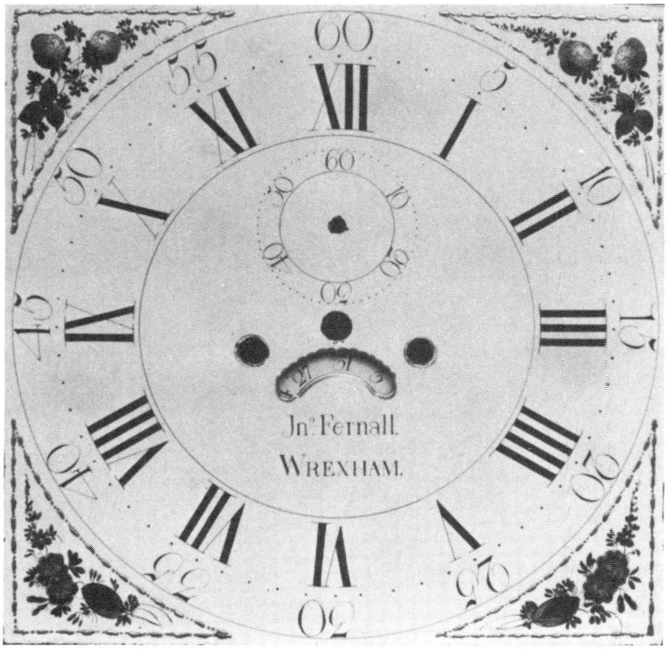

133 Restoration of Osborne dial complete. See page 211

Bartley dial resided in an attic for a great many years, and as the surface of the dial had become slightly porous, perhaps due to the washing activities, the heavy layer of dirt and other substances proved very hard to remove. However, eventually the dial turned out to be a most attractive one (**Illustration No. 135**) with gold leaf decorations in the corners, highlighted by yellow ochre and pink. The UV lamp was invaluable in finding out where all the very fine lines were in the corners.

Where the graphics are in near perfect condition, so called dry cleaned can be undertaken. This important oval dial made by Walker for Samuel Deacon of Barton in 1811 had a heavy deposit of atmospheric dirt with some oil staining around the winding holes (**Illustration No. 136**). The very thick varnish on the artwork was discoloured as well, and this varnish appeared to be cracking in places. After brushing off some of the surface dirt, 800 grade silicon carbide was used on cotton

---

134 A Walker and Finnemore dial *c.* 1810 that had been washed very thoroughly and then abandoned for many years. The corners are a Finnemore design. See pages 211 and 212

135 The dial restored and it can be seen to have been made for T. Bartley of Denbigh. The colours in the corners are pink and yellow with gold leaf and many fine lines. See page 212

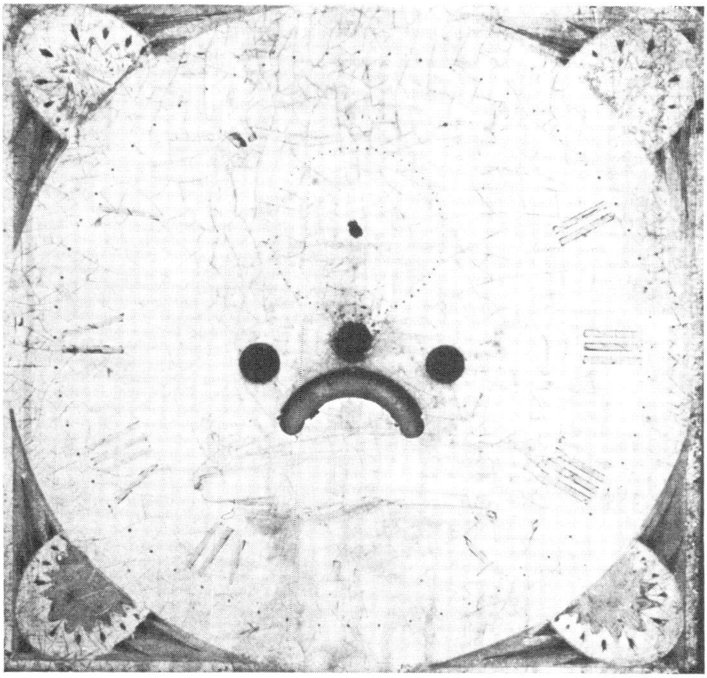

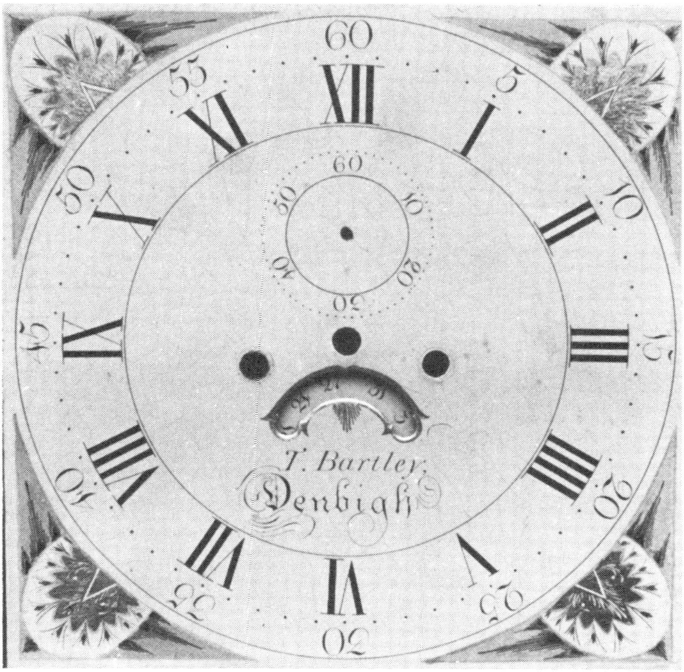
T. Bartley
Denbigh

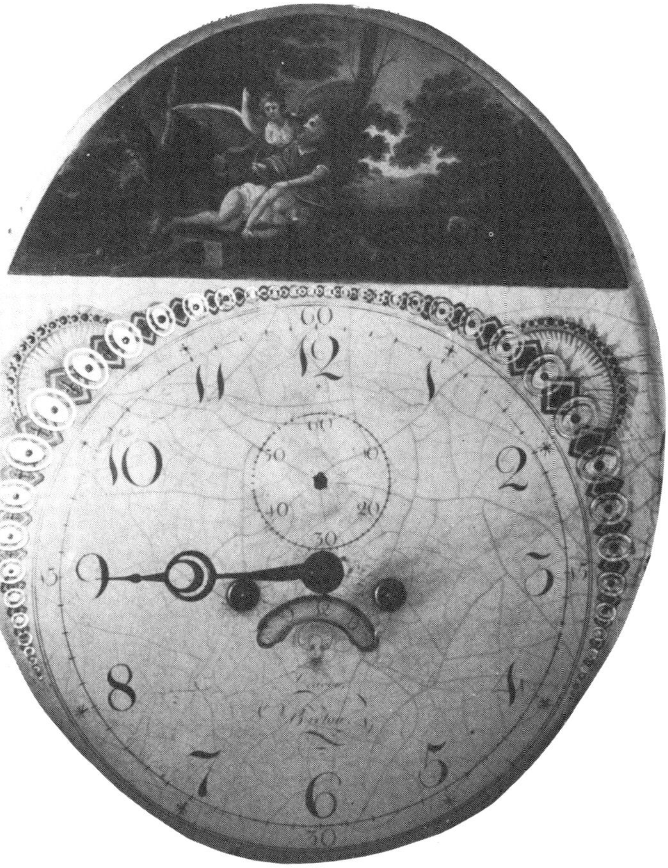

136 Atmospheric pollution and oil damage can be seen on this Walker dial which has a movement dated 1811. See page 212

flannel covered swabs employed in a rotary fashion. The swabs were dampened slightly with distilled water. An area of no more than 1″ or 2″ square was attempted at one time, cleaning off the area with a dry swab before the next section was commenced. This method takes a great deal of time so as not to remove any of the original graphics. Partial cleaning can be seen in **Illustration No. 137**) where a lot of the dirt has been removed, revealing a much paler ground colour underneath. Through the passage of time much of the very fine scrolling around the name had been lost, though the gold leaf work in the circles was in good condition. The whole dial, when all the dry cleaning was finished and the dark varnish removed from the break arch scene, was really remarkable. This break arch scene depicts the sacrifice of Isaac by Abraham and, although rather dark, a very worried looking ram can be

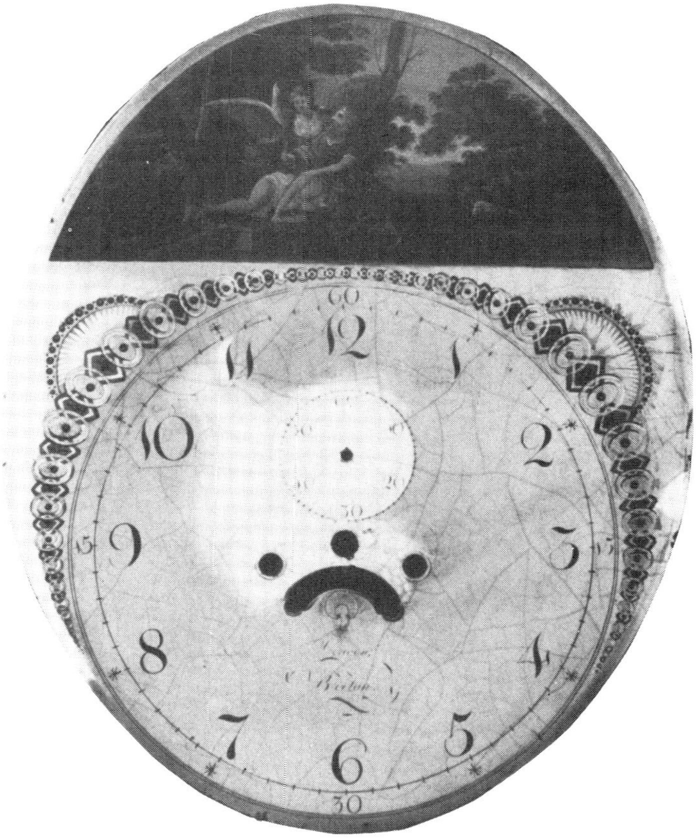

137 Partial dry cleaning of Walker dial. See page 214

just seen on the right. All of the missing scrolling around the name was put back with the help of the UV lamp, which also found the missing 'rays' in the corners (**Illustration No. 138**). The clock is on display at Newark Houses Museum, Leicester.

Rust

Rust is one of the most common problems with painted dials. It ranges in severity from the crumbling wreck (**Illustration No.6**) of the Birchall of Nantwich dial, which necessitated a complete strip and repaint (**Illustration No. 139**), to the minor damage often found at the bottom of a dial. It is sometimes difficult to decide how far to go when dealing with rust. Usually it is best to clean the dial first to see how much real rust damage there is. Anything really loose and crumbling has to be removed, either by scalpel or engraving point on the power tool. Some small bubbled up areas that have been

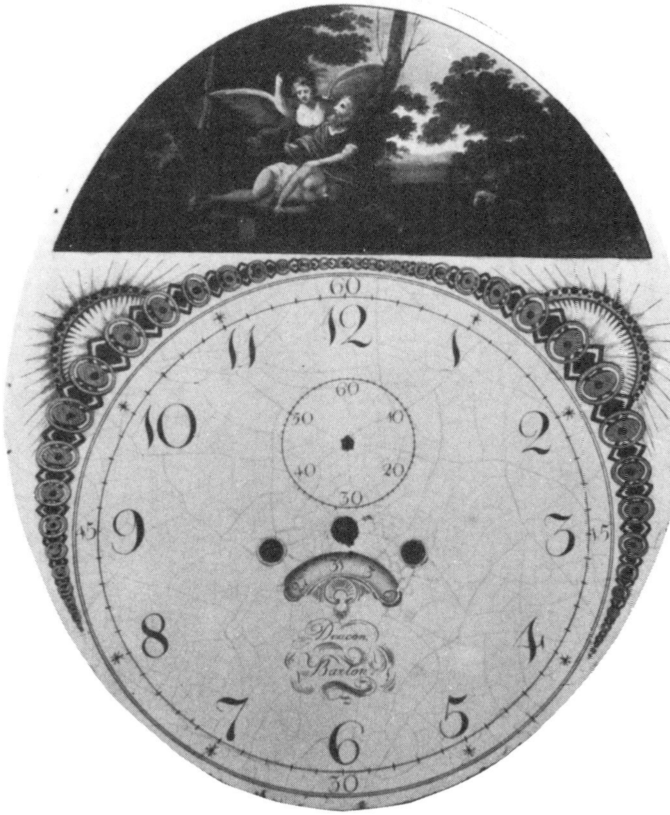

138 A very fine and unusual dial made for Samuel Deacon of Barton is revealed. The corners, which appear almost to revolve, are typical Walker work. The break arch scene is 'The Sacrifice of Isaac by Abraham'. See pages 214 and 215

there for years and have not spread can be a possible future risk, but it is probably best to leave these if the top of the blister has not cracked open. The mounted grinding point smooths off the very rough metal ready for the filler. A few words on the use of 'plastic padding' filler might come in useful here because in the case of really bad rust damage a great deal of filling must be done. Filling should be done a little at a time, especially in the case of large holes. It should be put on economically and smoothed off at once with a clean plastic straight edge normally supplied with the filler. The straight edge must be cleaned out with solvent (surgical spirit) at each use to prevent dragging. This filler can tend to separate out in the tube, producing a lot of oil, so it is best to mix this up in the tube before taking any out. The author has found a small rigid stainless steel spatula 1'' wide

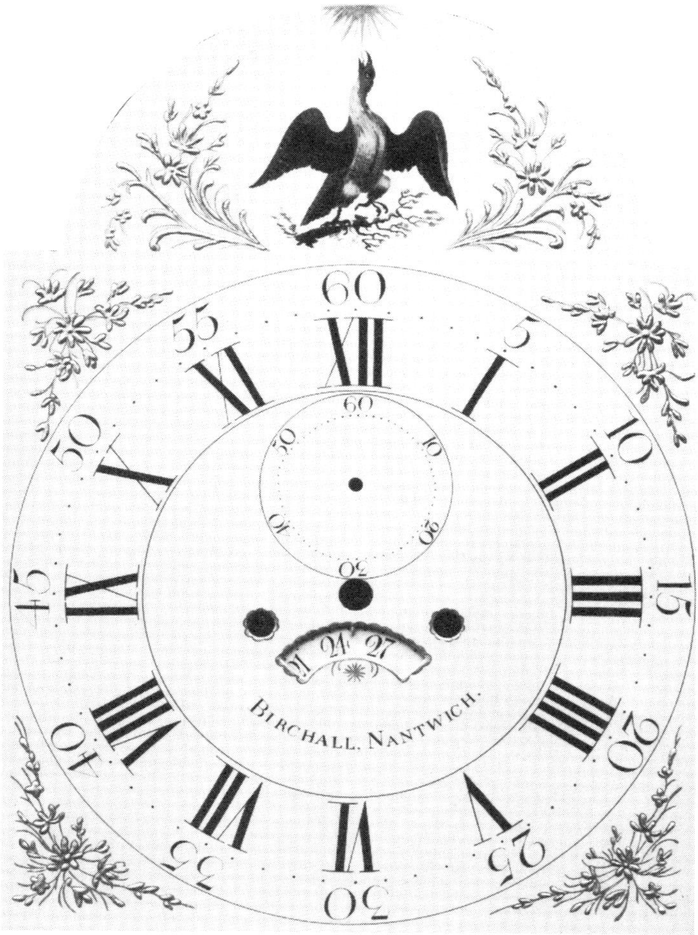

139 Very bad rust damage (see Illustration No. 6) necessitated a complete strip and repaint of this very fine Osborne and Wilson dial 1772–7 made for Birchall of Nantwich. See page 215

to be the best mixing tool, and a 3″ square of Formica or other laminated plastic provides a good surface to mix upon, as it can be easily cleaned with solvent after the mixture has 'gone off'. This, of course, depends greatly on the amount of hardener used, as well as the temperature. The quantity of hardener employed should just colour the grey base, and it must be mixed quickly and thoroughly. The spatula is useful for large areas, but small ones are better filled using a knife point.

After hardening about one hour, excess filler, where there is quite a lot, can be removed using a scalpel with extreme care. In the case of only a thin layer, hand pressure on a linen-covered swab, based on a stout piece of pegwood will take away a lot, though sometimes a small amount of solvent must be used. Obviously the less excess filler there is, the better, and any excess filler must be removed on the same day as the application, otherwise further hardening makes the work even more difficult. The idea is to build up a flat surface just below the level of the base paint. Where there is a distortion of the paint layer, this can be hard to do. The most difficult areas to fill are either on the edges of the dial where the paint layers may be thin, or on the edges of the date aperture. Files of various sorts can come in very useful in order to smooth off the surface in order to produce a rounded edge, and an old escapement file makes a good finisher in small areas. Emery boards such as used on the fingernails are useful on the long straight edges of a dial. Sometimes the fingertips can find a rough place that the eye does not see. After filling, the procedure is the same as that done for the delaminated dial.

Rust and Crumbling Gold Leaf
Rust can, and often is, combined with other damage such as oil, etc. A really horrible example is a dial that was left on an outhouse for years. It

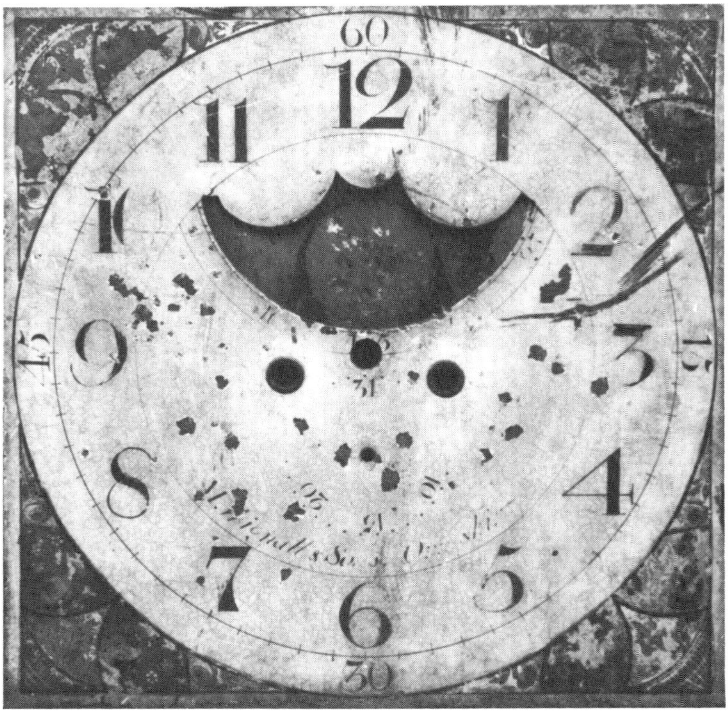

140 An abandoned and mistreated Whitaker and Shreve of Halifax dial which shows bullet marks, subsequent rust, dirt and oil as well as crumbling gold leaf corners. See page 219

had rust, oil damage, crumbling gold and silver leaf work and it had also been used for target practice (**Illustration No. 140**). The brass moon had literally almost been shot to pieces, though luckily none of the shot holes actually went through the metal (**Illustration No. 141**). This dial had been made by Whitaker and Shreve of Halifax and had the still recognizable D26 on the back of the main dial. The decorated corners (**Illustration No. 142**) crumbled off at a touch, but originally they had the green and gold effect. Although the graphics were still there they had to be removed because of the amount of damage together with all the dirt and grease. Rust had started in the shot holes, which can clearly be seen in the partially cleaned areas (**Illustration No. 143**). The moon dial was also partially cleaned in this photograph. IMS and pumice powder was used in the cleaning as well as some scalpel work in the corners. It took a long time and an entire single bed linen sheet to obtain a sound surface. The ghost images of numerals, name and even maps

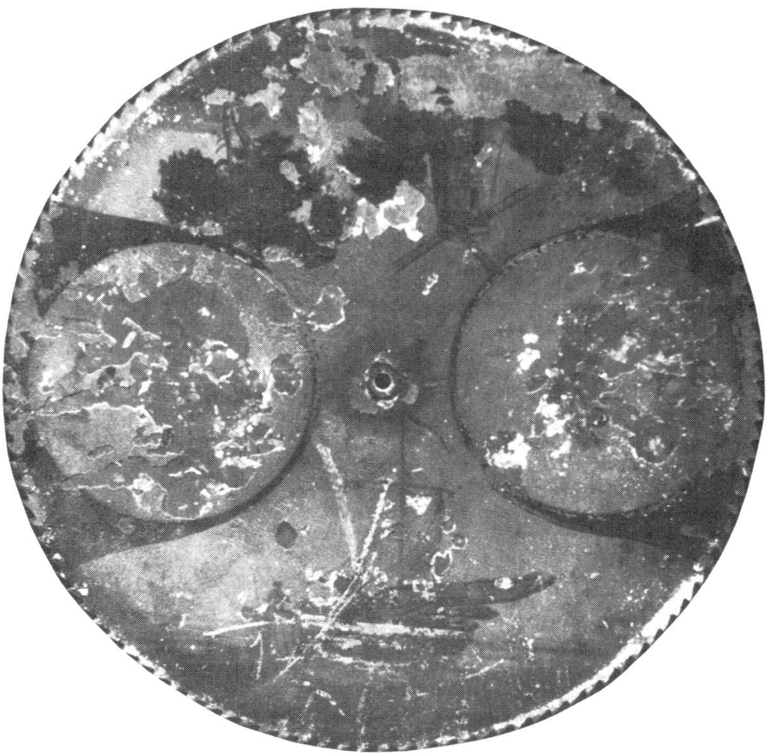

141 The brass moon dial to Illustration No. 140 which has been extensively damaged by use as a practice target. See pages 219–20

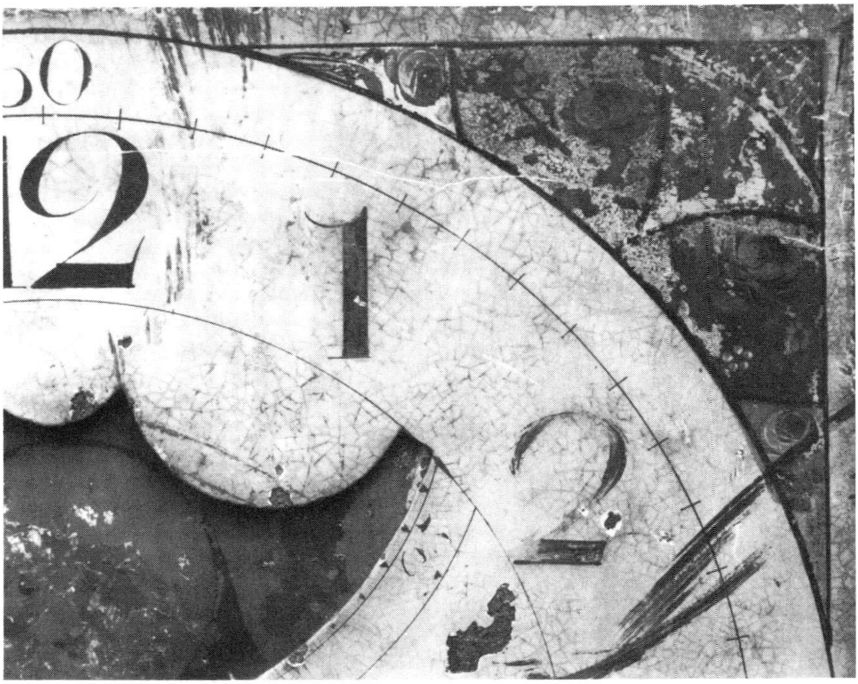

142 A detail of the Halifax dial showing the crumbling of a corner before restoration. See page 219

could just be seen once the dial was completely cleaned, filled and coloured in (**Illustration No. 144**). Parts of the roses in the corners could be saved, but most of the corners had to be regilded. It was decided to use gold leaf for everything, and not attempt the more unstable silver leaf under the green translucent paint. Eventually what emerged was a very nice dial indeed (**Illustration No. 145**) made for Mary Wignall and Sons, *c.* 1803. The moon had to have a large amount of inpainting, but also the worst bulges in the brass needed flattening out. This moon is typical Whitaker and Shreve work of that period where, luckily, the brass used was of good quality (**Illustration No. 146**). The dial as a whole had a combination of inbuilt problems (silver leaf), accidental and wilful damage. The author hopes that there are not too many examples of this sort of thing, for the restoration took nearly a year. However, there was an immense amount of satisfaction

143 Partially cleaned Halifax dial showing rust in bare metal patches. See page 219

144 Cleaned and repaired Halifax dial showing ghost images of graphics and replacement of gold leaf in the corners. See page 220

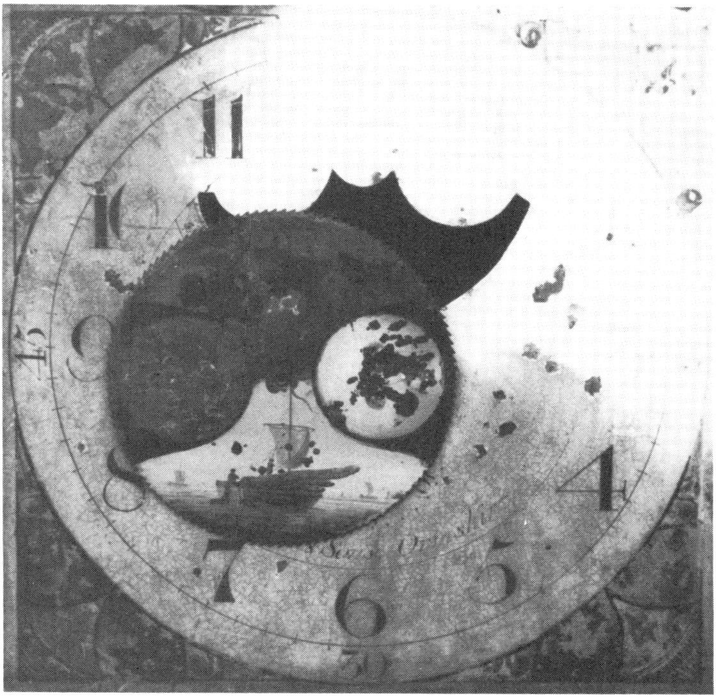

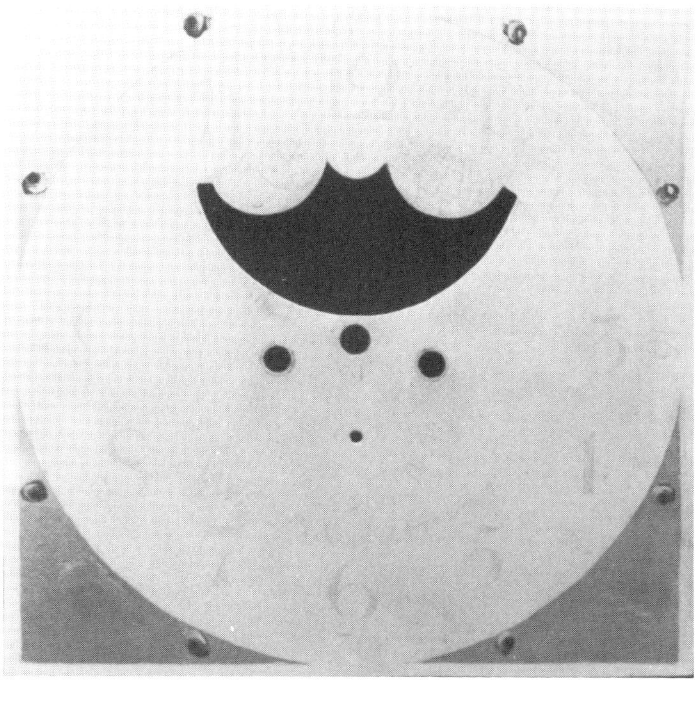

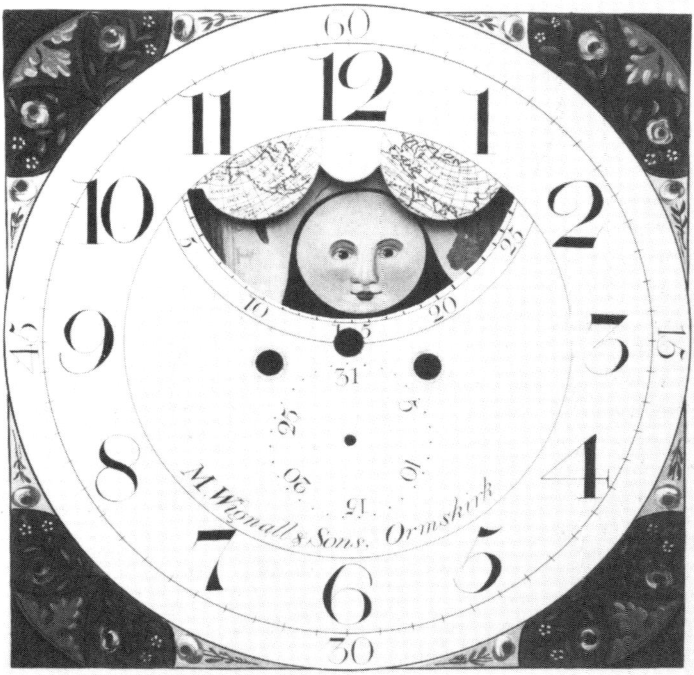
M. Wignall & Sons. Ormskirk

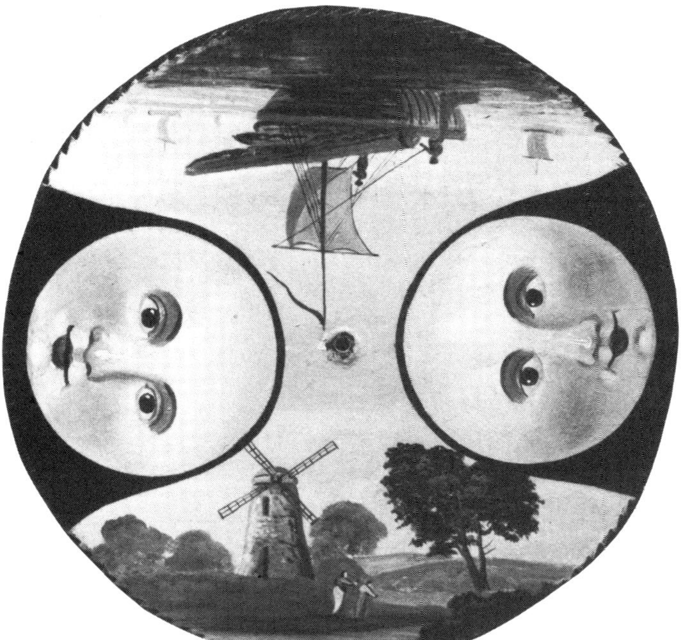

gained in rescuing such a piece from oblivion and eventual disintegration.

Rust and Delamination

Rust damage can be combined with bad flaking off of the paint due to delamination or other causes. It may be necessary to remove a whole dial centre which contained all of the graphics, retaining the painted break arch and corner work. More difficult is the removal of the chapter ring only in the case of a dial with a fully painted dial centre. An extreme case is the removal of all of the square portion of the dial, leaving only the break arch. It is a very drastic step to take, but at least the original artwork can usually be saved. Naturally all of the graphics must be traced as well as the artwork. Most of the dials which need this treatment are post-1800 ones. The dial where only the break arch was left had no primer at all on it, and was a Halifax Newcastle one, 1810–15. The procedure used for the replacing of a dial centre will be described, though the steps are essentially the same for any work of this type. The first cut should be made with a sharp knife just beyond the chapter ring area, as it is easier to join up old and new work where there is some painted decoration. Obviously no chemical stripper can be used, but a good sharp scraper and some scalpel work should remove all of the insecure flaking paint. If the rust was very bad, it is a good idea to use a rust inhibitor (usually based on phosphoric acid), being very careful this does not get onto the painted areas. Then comes a coat of iron oxide paint, followed by a levelling skim of filler around the edges to make the transition from old to new less obvious. The iron oxide paint, being quite thick, is ideal in that it adds to the overall depth of the base paint. Any deep pits or holes can be filled in after using the iron oxide paint. After this comes the application of many coats of base paint, carefully sanding down between each coat when dry. Humbrol Enamel Matt, suitably coloured to match the original, is used for this until a satisfactory surface has been obtained. Depending on the finish of the original paint remaining, the final coat can be matt, polished with 800 grade silica carbide, or gloss, matted down with the same powder. When the base paint is really dry the graphics can be put back, and inpainting done to help cover the join.

A word about tracing and transferring numerals might be applicable here. Of course Roman numerals can be pricked out, but Arabic ones

---

145 Restored Halifax dial, made for Mary Wignall and Sons, of Ormskirk *c.* 1803. See page 220

146 Restored moon dial to the Wignall clock. See page 220

are more difficult. It goes without saying that carbon paper must not be used, as it is opaque and makes a mess. The best system is to lightly go over the back of the tracing with a 4B pencil, used sideways so that the whole numeral to be traced is covered with a thin layer of graphite. Then the paper must be positioned and firmly held while going over the numeral with a 4 or 5H pencil, being careful not to press down with the fingers which will leave graphite where one does not want it. This simple information is included because a student became most distressed when carbon paper left very smudgy marks. As most of the dials that need this treatment are later ones, they usually have a gold leaf band around the edge of the outside of the chapter ring. The gold leaf will show up any irregularities on the surface, so it is especially important that this area is really flat. The artwork must be thoroughly dry before putting on the gold leaf, otherwise the leaf will stick where it is not wanted.

Dropped Dials

The dropped or knocked over dial can be badly bent, with loose feet and cracks all over. Some straightening can be carried out with care, especially if it is a corner or a moon arch, but too much bending back will result in a lot of paint coming off. Loose feet must be re-rivetted, and if no false plate was used the front plate of the movement should be available to check that the feet still fit. Sometimes whole new feet have to be made because they have fallen off and disappeared, but where one old foot remains the new feet should follow the pattern of the old one. Moon and date posts can also be bent or become loose and it is essential to check this before any restoration is attempted. Moon and date dials can get twisted as well, and the centre fitting of these may have to be re-done.

If the accident happened a long time ago, and someone attempted to straighten the dial, plier and vice marks are often seen. The accident of falling over seems to have been a common one, yet a great number of clock cases have numerous holes in the back board which would indicate that the clock had been attached to the wall.

Fingernail Scratches

There is another type of damage which, though accidental, is really manmade through carelessness. The dreaded fingernail scratch caused by inattentive winding or moving the hands around can really make a mess, especially of a newly restored dial. Fingernails are very hard, and they transfer black pigment from the numerals to the plain areas in a clockwise direction. They can also occur in old, completely unrestored dials. Frequently these scratches are seen at the bottom of the dial. Sometimes, if not too bad, the marks can be removed with careful use of the silicon carbide 800 grade. An extreme example of fingernail scratch is

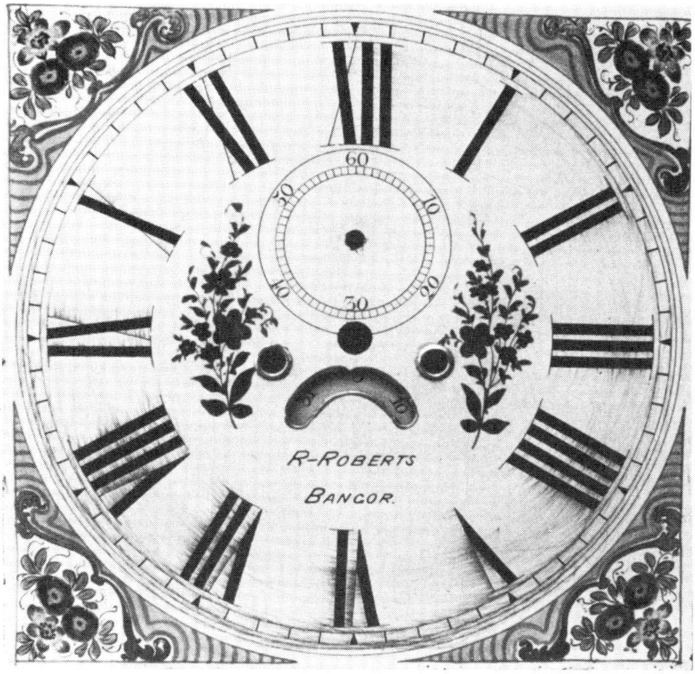

147 The results of fingernail scratch on a Fletcher dial made for Robert Roberts *c.* 1845. See pages 224–5

on a dial made by John Wright Fletcher for Robert Roberts of Bangor (1828–56) (**Illustration No. 147**), where the damage was so bad that the dial had to be completely re-done, emphasizing the fact that all restoration must be reversible. This problem of fingernail scratch cannot be put too strongly, as both curatorial staff and private owners may not realize what can happen. All the notes accompanying dials emphasize this point.

DELIBERATE MANMADE DAMAGE, OR GETTING RID OF FATHER TIME

Overpainting of Numerals

Manmade damage seems to happen very frequently to painted dials. In fact it is rare that a restorer can see a dial that has not been touched and has not got anything else the matter with it. Overpainting of the numerals is the most common, usually because of previous washing or even scouring. Frequently a dial has the sort of overpainting on the graphics that just shouts 'wrong', like this Wilson one (**Illustration No. 148**) which was made for John Clifton of Liverpool (courtesy of Liverpool Museum) sometime in the mid-1780s. The maps are a

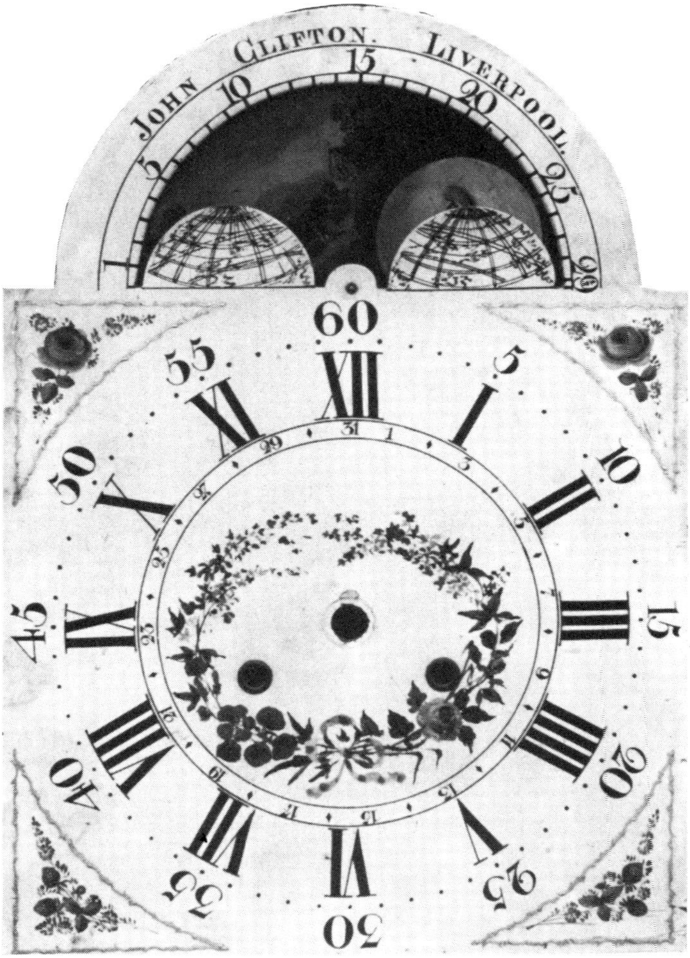

148 Overpainted graphics including maps on this Wilson dial *c.* 1785 that was made for John Clifton of Liverpool. See page 226

cartographer's nightmare and the Arabic numerals far too heavy. After removing the dirt, grease and overpainting of the graphics, the original form of these could be seen under UV light (**Illustration No. 149**). The crazy white lines on the maps are the previous improver's idea of latitude and longitude while the darker lines are the correct original ones that have been pencilled in. These maps are the normal Wilson ones of the 1780s. The fat overpainted Arabic numerals show up as pale grey, while the original more elegant and slender ones are a bright white. Even the Roman numerals are slightly incorrect, as seen in the top of the XII. This dial restored very well (**Illustration No. 150**) and was probably

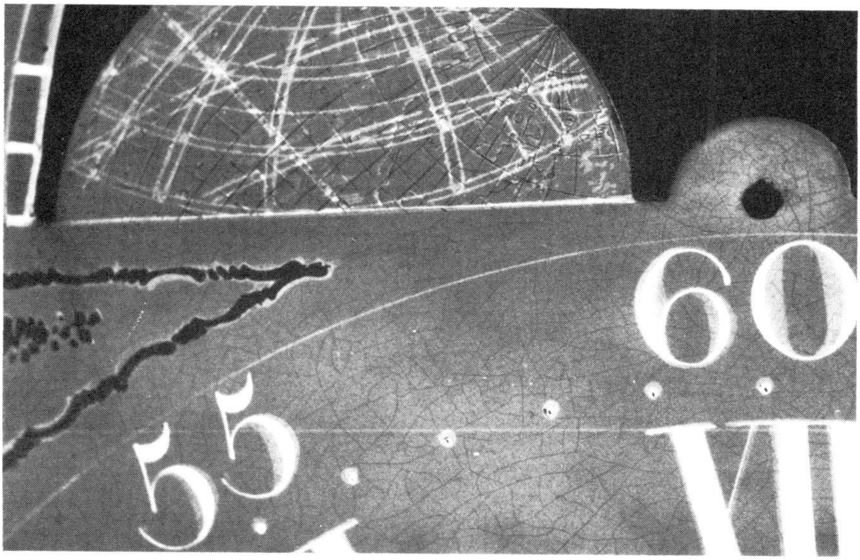

149 The cleaned Clifton dial under UV light showing original graphics. The overpainting of the numerals is pale grey except for the maps where the original faint latitude and longitude lines have been pencilled in by the author. This shows up as much darker and thinner than the spurious markings. See page 226

painted by the same artist as the 'Beatrix Potter' dial (**Illustration No. 21**) as the gesso work and the lower corners are almost identical. Garlands in the centre of a Wilson dial are rather unusual, however.

UV photography is not normally expected by a customer, but it can be very useful before any restoration is done, in order to convince the owner that some of the previous graphic work was not original. It can be almost impossible to see the difference between overpainting and original work by normal light. This UV photograph (**Illustration No. 151**) is a detail of the bracket clock made by Barker of Wigan and Warrington (seen in **Colour Illustration (xix)**). The number 5 can be seen to be different from the original, as well as the X. This brings up a point that is not always well understood, and should be repeated here. Usually before 1820 the thin slanting lines of the V + X in Roman numerals ended in the plain serif, unemphasized by a thickening at the juncture of the two lines. Many quite reasonably restored Wilson dials, as well as others, are marred by this obvious defect. If a UV lamp had been used it would have been evident what the shape of the numeral really was.

Frequently the scrubbing operations have been all too successful, especially in the case of rolled on transfer decoration which, because of the type of ink used, was easily removed. This Osborne dial

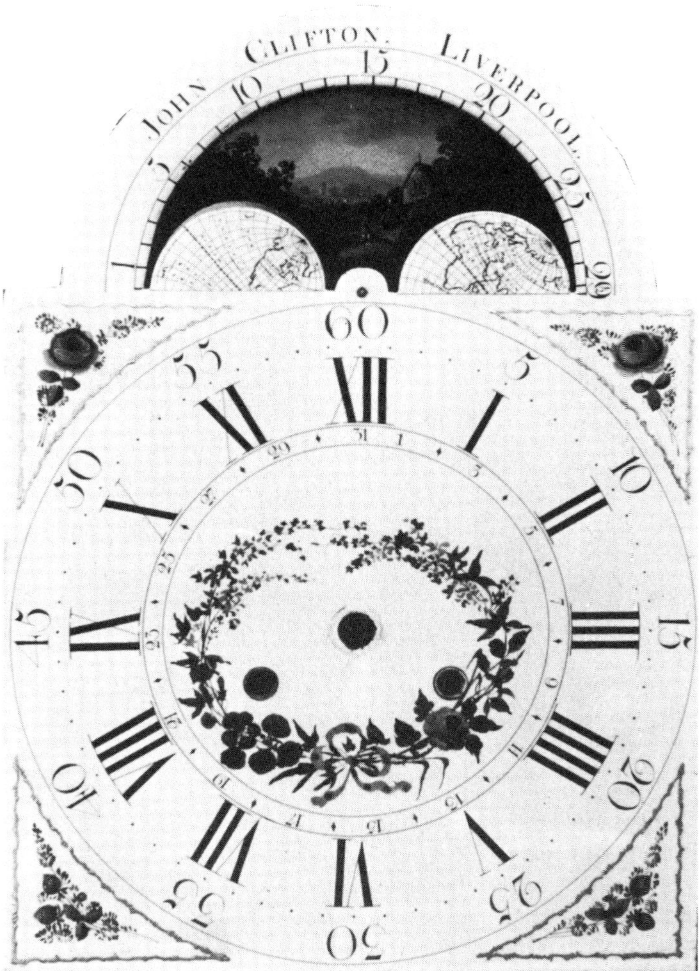

150 The fully restored Clifton dial showing the early Wilson cut-off style maps and the fine gesso framing the corners. See page 227

(Illustration No. 152) was a sad sight when it first arrived, giving no clue at all to the naked eye as to what had been on the corners and break arch. The remains of the graphics were there, but when this dial was put under UV light all the missing work was revealed. It took many days to put back the decorative work using a pencil on the dial very lightly recording what the lamp showed up, and after the retouching varnish was put on the laborious inking in was done, using the finest mapping or crow quill pen and Rötring ink. As is true with map work, such details must be done as soon as possible after the varnish is dry to the light touch, because the varnish can become brittle and fracture under

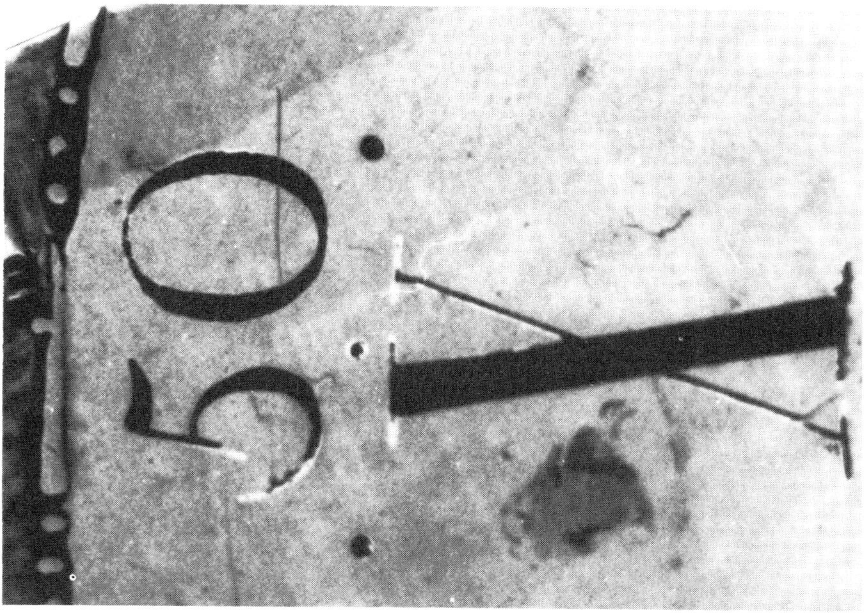

151 UV photograph of Barker bracket clock dial (Colour Illustration No. *xix*) showing previous restoration and original graphics. See pages 120 and 227

pressure of the steel nib. The finished dial (**Illustration No. 153**) was a great surprise to the owners. Thomas Bayley of Burton (*c.* 1790) is listed by Loomes as being pre-1764 and the Burton located in Westmoreland, but certainly this was a different Thomas Bayley and the Burton was the Cheshire one. The compass rose in the centre of the seconds ring has been seen on quite a few other Cheshire clocks. One other similar Thomas Bayley transfer print dial is known to exist somewhere in North Wales. These all transfer print decorated dials are rare, and most of them appear to have been made in the 1785–90 period. They are not always recognized for what they are, and need very careful restoration.

This section was headed 'Getting Rid of Father Time', and he was painted by 'FW' in 1982 (**Illustration No. 154**), using a felt tipped pen. It is not certain if the Roman numerals were done by the same person or not. The dial, besides being heavily scoured, was also further defaced by the erroneous addition of a seconds ring and a large hole where the original name had been. The movement had no provision for a seconds hand when it was made. The dial surface was very dry, but luckily all traces of Father Time vanished except under the UV lamp. The dial turned out to be a typical Walker and Hughes one (**Illustration No. 155**), made for Samuel Smith of Walton (on Trent) (1835–49). The delicate shell corners luckily had enough remaining to see what they had

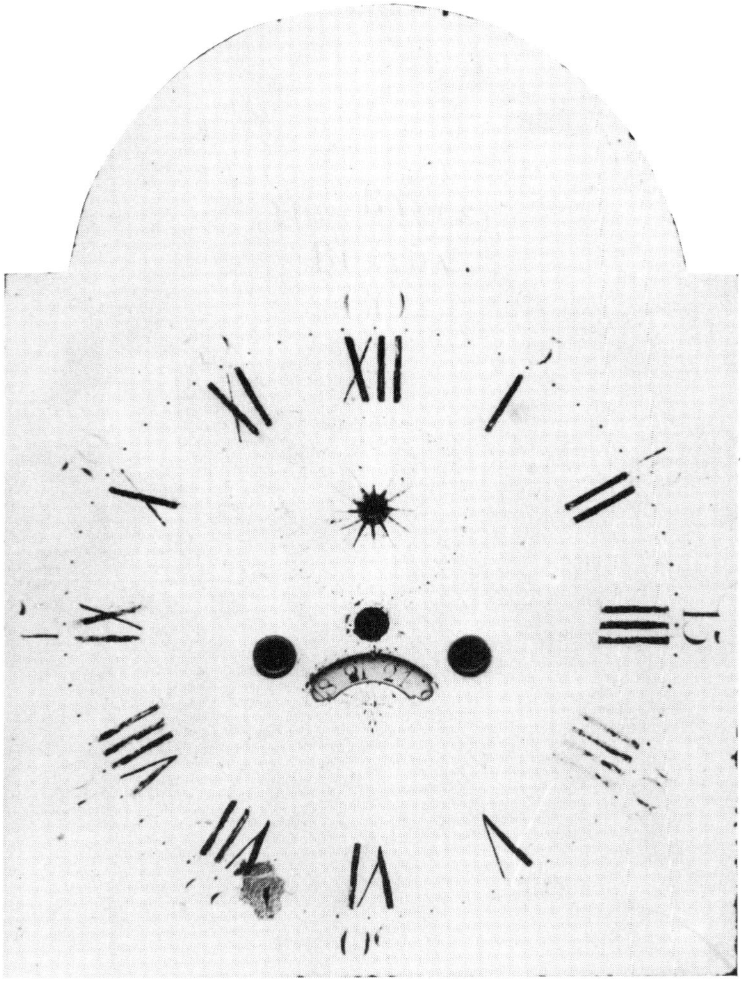

152 An Osborne dial with all of the transfer print decorations washed off and Roman numerals clumsily repainted. See pages 227 and 228

been, but the dog tooth gold leaf was only left in the lower right hand corner.

Sometimes repainted graphics can be quite good, as in this example (**Illustration No. 156**), but the flowers in the centre of the late Wilson dial somehow do not look quite right. Both flowers and name came off with one wipe of solvent, revealing a quite different name and town, plus two full rigged ships underneath (**Illustration No. 157**). The name Calcott was afterwards seen to have been cast into the bell. John Calcott (1776–1830) is listed as making clocks and watches, but he might have

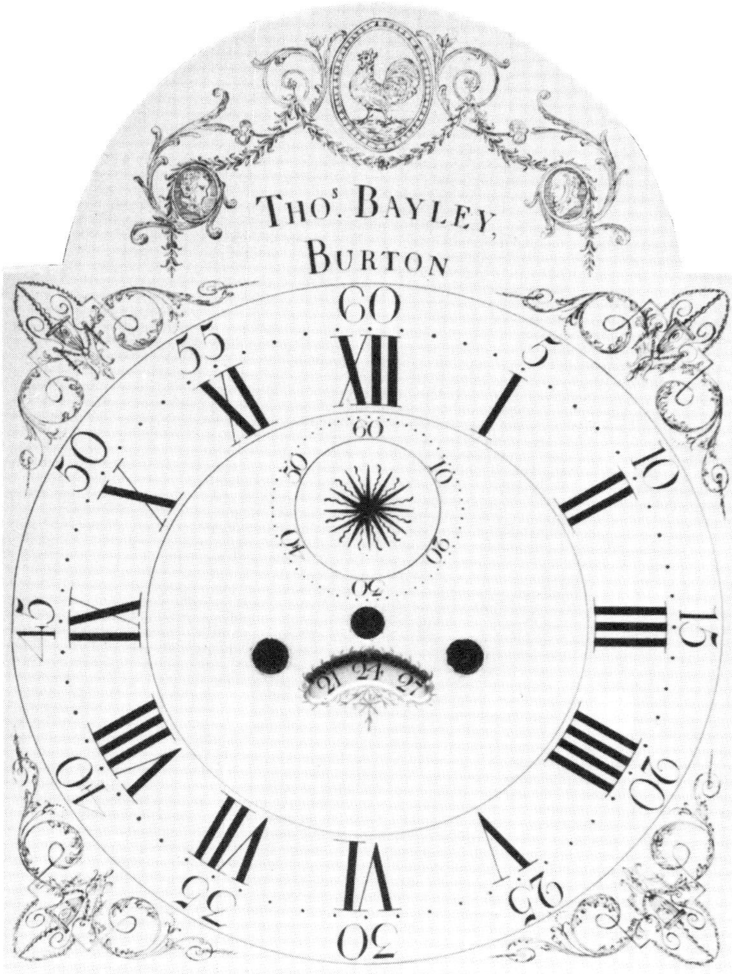

153 Restoration completed and the clock maker's name, Thomas Bayley of Burton, Cheshire 1785–90 now revealed. These corners are found on other dials but so far the crowing cock in the break arch is unique. The two profiles are Plato and Socrates. See page 229

made bells as well. This type of ship has turned up on other Wilson dials of the 1803–9 period.

Overpainting of Whole Dial Surface

The complete overpainting of a whole dial is, luckily, much more rare, but it certainly does happen, and this pathetic object (**Illustration No. 158**) had been overpainted twice, each layer requiring a different solvent. The top layer came off with acetone, and the bottom one needed surgical spirit. The dial had been made by Richard Hipkiss for Lawrence Shakeshaft of Preston about 1810, and the corners were very

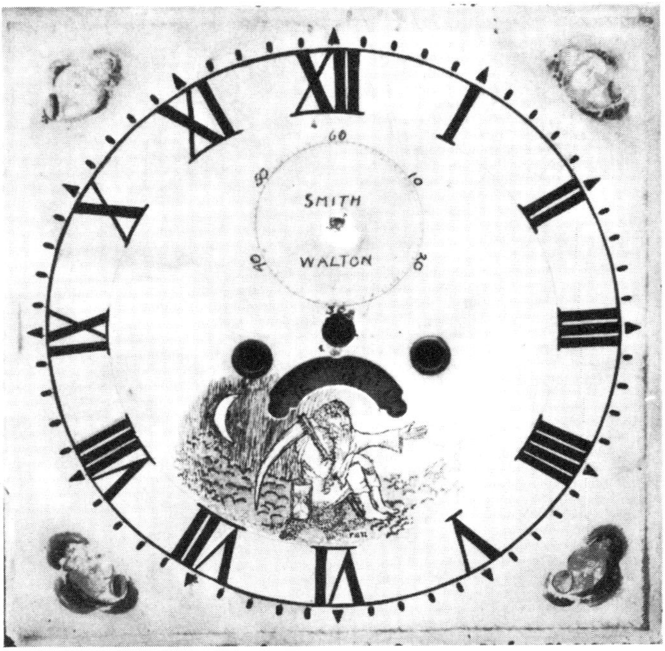
SMITH
WALTON

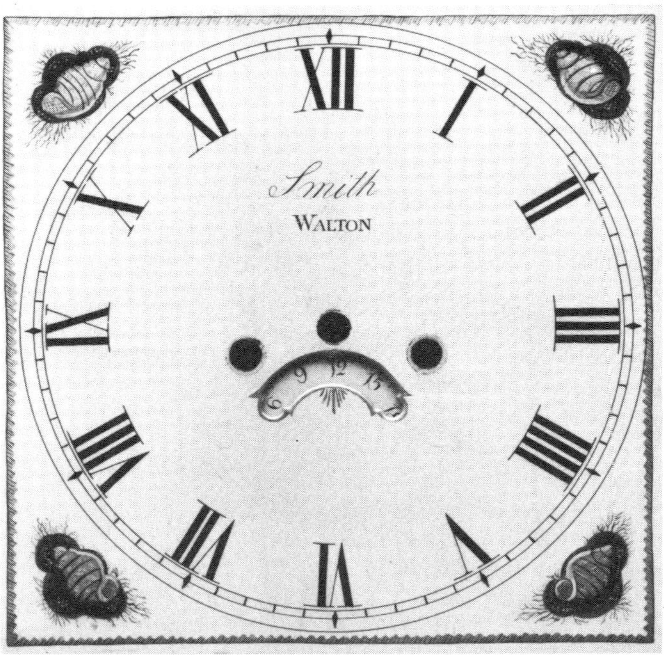
Smith
WALTON

brilliant in green, rose and gold leaf (**Illustration No. 159**). Sometimes with this heavy sort of overpainting it is necessary to mix the solvent with some pumice powder, and judicious use of the scalpel blade also helps to remove the softened layer, but great care must be taken not to damage anything underneath.

Oiling

Oil, as mentioned before, can do a lot of damage. One of the really bad cases of this sort can be seen in this Walker and Hughes dial *c.* 1820 made for Peter Evers of Chester (1791–1828) (**Illustration No. 160**). Some attempt had been made to wash the dial a long time ago, but the oil had gone just about everywhere, and the long vertical streak on the right hand side was oil. Repeated de-greasing was necessary, using 'Genclene', and the application of orris root powder with gentle heat helped to bring out a lot of the oil in the cracks, but the paint around the centre hole had softened and crumbled, so quite a lot of that had to be replaced. The oil had even attacked parts of the artwork near to where the gold leaf on the ladies' dresses were on the left hand side. Also, the varnish on all the artwork had darkened a great deal, especially on the moon dial. Sometimes it is impossible to remove all of the oil stains, and some residual staining is left, but in this case the staining was minimal (**Illustration No. 161**). This dial was the author's first commercial attempt at restoration, before the purchase of the UV lamp, so the maps lack some detail.

There is a type of damage caused by a form of wax or oil contained in furniture polish which has been carelessly applied to the mask of the hood so that the edge of the dial is contaminated. Normally the mask was left unpolished, but there have been cases where it was even varnished, with the dial left in situ. The furniture polish leaves an almost indelible brown stain all around the edges of the dial (**Illustration No. 162**). There had been some attempt in the past to wash this dial and put the Roman numerals back, and the marks of the cloth can be seen clearly on the left hand side. The dial is a fine Wilson one, made for George Clapham of Brigg *c.* 1785. It was almost certainly painted by the

---

154 This is an example of a dreadful dial desecration. The original name of the clock maker had been drilled for a seconds hand not included in the movement. The whole dial was heavily scrubbed and incredibly bad graphics put on together with 'Father Time' done in felt-tipped pen in 1982. See page 229

155 A very attractive nice Walker and Hughes dial *c.* 1825 with no trace of 'Father Time'. See pages 229–30

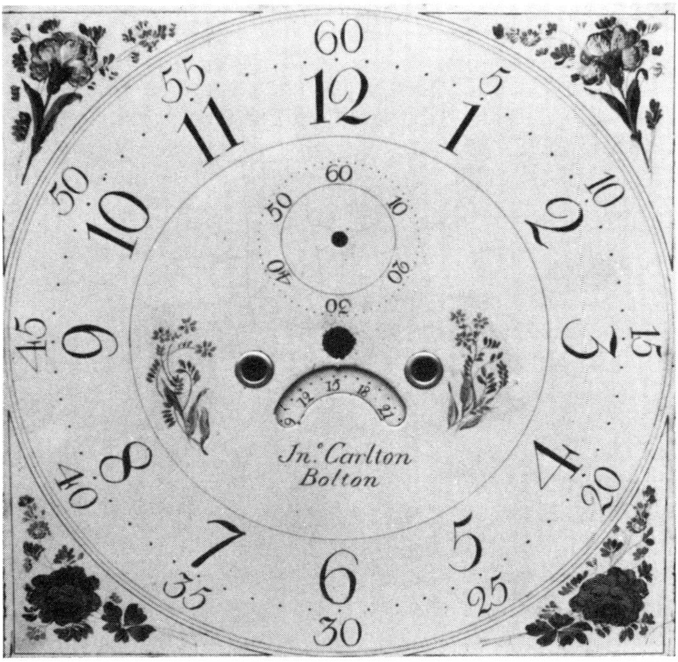
Jno Carlton
Bolton

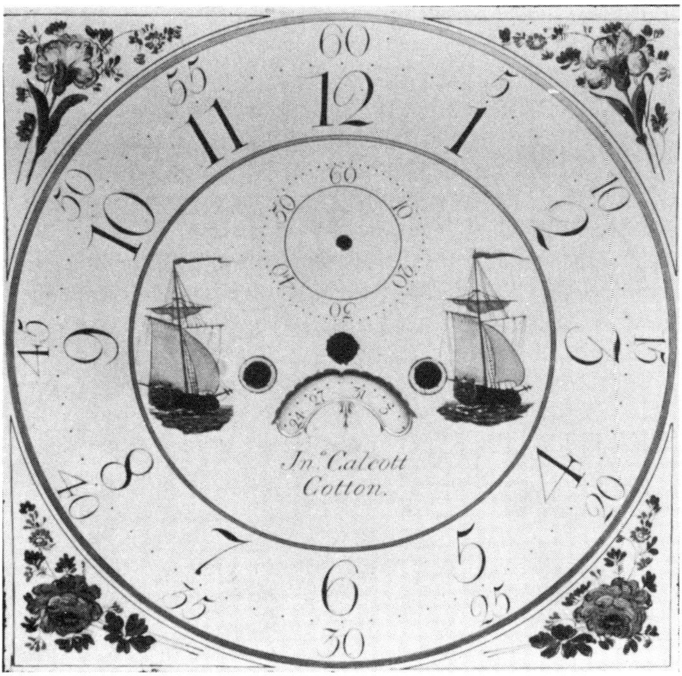
Jno Calcott
Cotton.

same artist as the Beatrix Potter (**Illustration No. 21**) and John Clifton (**Illustration No. 150**) dials, though the transfer printed decorations in the break arch have not been seen before. The fully restored dial (**Illustration No. 163**) shows a good example of restrained elegance typical of this period.

MAKING A NEW DIAL, OR A COMPLETE REPAINT

Doing a complete repaint means a lot of preliminary work, such as photography, tracing, etc. It is obviously wiser to wait until the photographs have been developed or printed before stripping the paint off. If there is no date dial with base paint on it, some flakes of this main dial should be kept for colour matching. When it is deemed necessary to keep the back of the dial in its original condition, no stripper, of course, can be used, and there is a lot of hand work with a scraper and scalpel. A new dial gives one somewhat more scope, but many new dials are not a success because the basic rules of style and proportion have not been adhered to. It is always best to base a new dial on one that has already existed, and if it is for an extant or existing case the style of the dial must be of the correct period for that case. Any new dial containing automata has to be very carefully figured out with regard to the amount of movement and its method of attachment. A plan of a new dial based on an old one can easily be made using a good enlarger and a colour slide of an original dial. Cartridge paper should be firmly glued to the smooth side of a piece of hardboard using photographic mounting spray. It can take some time, but the graphics should be clearly pencilled in as well as the decorative details. This should be gone over in ink afterwards, correcting any obvious faults such as the chapter ring too far to one side etc., and then a working tracing can be made.

When dealing with a repaint on the original metal base, it is of course necessary to make sure there is a clean, rust-free surface where any holes or dips have been filled in with 'Plastic Padding'. A new dial also may have some depressions where the feet have been rivetted in. As mentioned before, iron oxide paint provides the ideal primer, and if an old dial plate is very pitted with small holes difficult to fill, several coats of this primer, duly sanded down, will help to provide a flat surface. Most people think old dials to be absolutely flat, but this was often not

---

156 Good repainted graphics but the name and town are wrong on this late Wilson dial. Note the ungraceful flowers in the dial centre. See page 230
157 Emergence of two ships hidden under the spurious flowers were quite a surprise to the owner! The name Calcott (of Cotton) was cast into the bell. The dial is *c.* 1806. See page 230

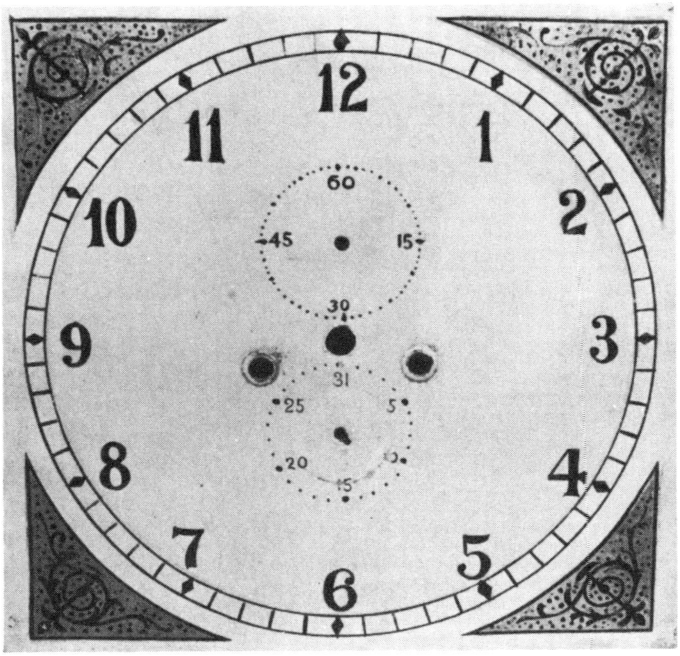
12
1
2
3
4
5
6
7
8
9
10
11
60
15
30
45
31
5
10
15
20
25

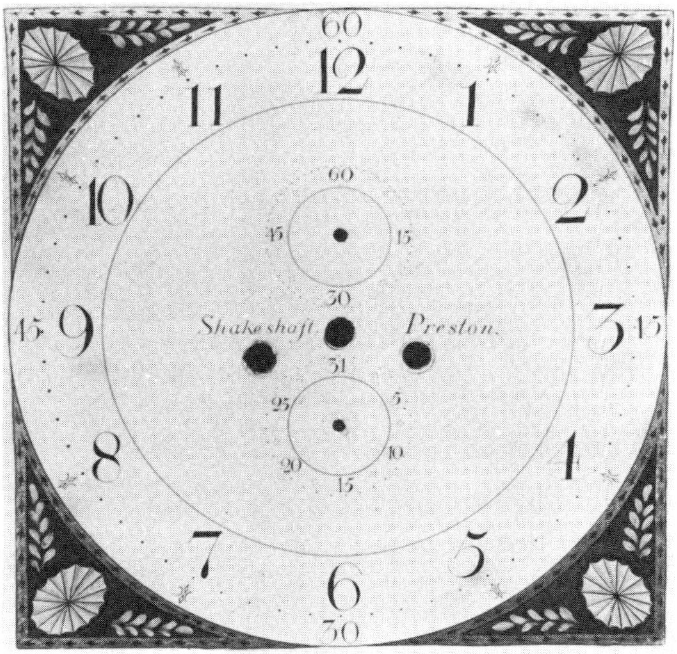
60
12
1
2
3
4
5
6
7
8
9
10
11
45
15
30
Shakeshaft. Preston.
31
5
10
15
20
25

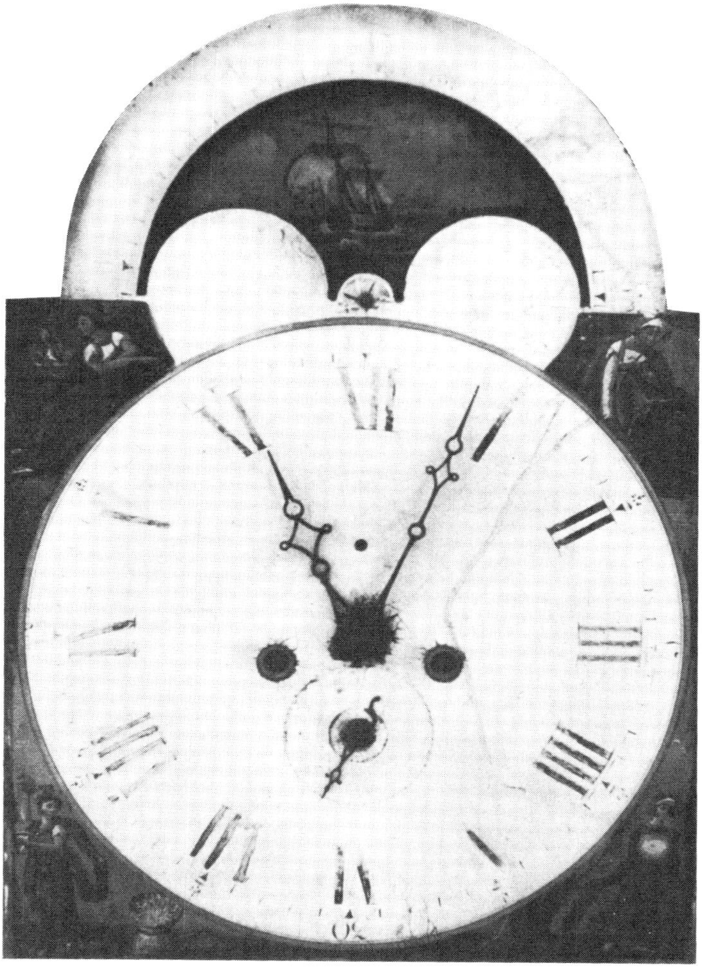

160 A very badly oil damaged Walker and Hughes dial *c.* 1820 with maps that have almost disappeared. See page 233

the case, so providing a flat surface can be somewhat of a compromise. Once all the primer is absolutely dry, the base painting can begin, using Humbrol Enamel Matt of roughly the same colour as the finished dial and a good 2′′ bristle brush. A word about brushes might be useful here.

---

158 This unfortunate Hipkiss dial had been overpainted twice and the clock maker's name completely lost. See page 231

159 After a great deal of work the original dial re-appears having been made for Lawrence Shakeshaft of Preston *c.* 1810. The corners are very bright, using gold leaf together with green and rose oil paint. See pages 231 and 233

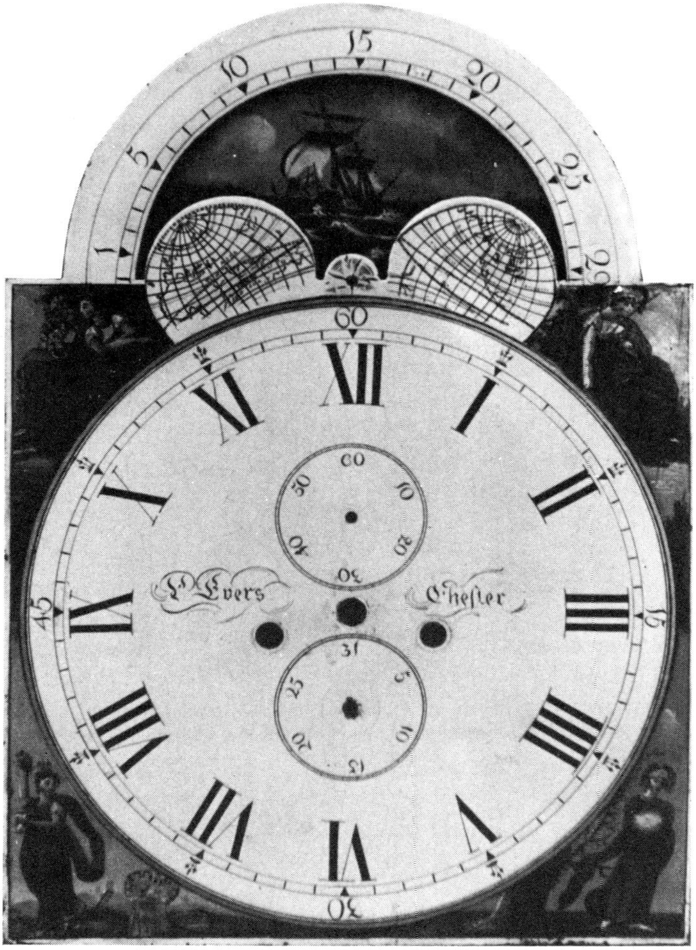

161 The colourful four seasons dial was made for Peter Evers of Chester using gold leaf under transclucent colours of the ladies' dresses. The Walker and Hughes trident motif is evident in the minute band. See page 233

Iron oxide is almost impossible to remove completely from a brush, so any brush used for this paint should be kept for this purpose only. Brushes should be cleaned out in white spirit immediately they are finished with, otherwise this paint quickly hardens. A solution of soap powder and very hot water also gets rid of paint in a brush, but it must be very thoroughly rinsed and dried before using again. A new brush must be used for the final coat of paint.

The painting should be done in as dust-free a room as possible and the worker's clothing must not shed hairs or fluff. It is unwise to paint on a very windy day, as the draughts stir up dust. A very hot day is also not good, as the paint dries too quickly. Paint applied too thickly will never really harden and tends to wrinkle at the edges. The paint itself may be too thick to even brush out properly, so some Humbrol thinner should be added. The paint must be completely dry before sanding down with wet and dry paper, using a cork block. The 250 grade is useful for preliminary work but it is very coarse, and 400 is the most practical, finishing off with the 800 grade for the last coats before the final one. About five coats of paint are usually necessary to form a good base, and the last two coats should be of the Humbrol Gloss paint. The first coat of gloss must be dulled down with the 800 grade paper, and the final coat should be flatted down with the 800 grade silicon carbide powder so that the ink and paint will 'take'. Sometimes it is advisable to go over the dial with 'Genclene' in order to remove any greasy fingerprints, because ink will not go on where there is any grease. Inevitably paint has run over the edges of the dial to form heavy bumps which should be removed and the tacky underneath allowed to dry out. Winding and other holes should be cleaned out to their original size, using a small file. The treatment of the date aperture depends on the era of the dial. Early dials had the date aperture neatly filed out before the gold leaf was put on, but later ones, after 1810 or so, often allowed the paint to remain. All of the base painting will take quite a few months as the paint must be really hard before undertaking the next step.

The dial must be very firmly attached to the turntable with some sort of sticky pads so that it cannot move. Then the tracing is lightly gone over on the wrong side with a 4 or 6B pencil used flat on all the numerals, name and decorations. The tracing is next put right way up over the dial, securing with sticky tape. It is important not to press down on the tracing as this will transfer some of the graphite to the dial surface. A scriber is used to prick out the Roman numerals in just the way old dials were done. Concentric circles can be marked by pricking or using a very hard pencil at points 12, 3, 6 and 9 respectively. Arabic numerals can be transferred by using a 6H pencil. Outlines for artwork are done in the same way. When one is sure all the details have been marked the paper can be removed. As one removes the paper, it is best to lightly pencil in the Roman numerals, because all these dots can be confusing. Artwork must be gone over in pencil.

Next the dial is removed from the turntable and placed on the frame stand. The centre hole locator is used to find the true centre according to the scriber marks. Often with old dials the centre that was used is definitely not the middle of the centre hole. The full circles should be done first and then the tops and bottoms of the Roman numerals. If there is to be a gilded band, this should be done with the pencil end to the compass so that the gold leaf can be applied after all the marking out is done, and when thoroughly dry and varnished the circles can be gone

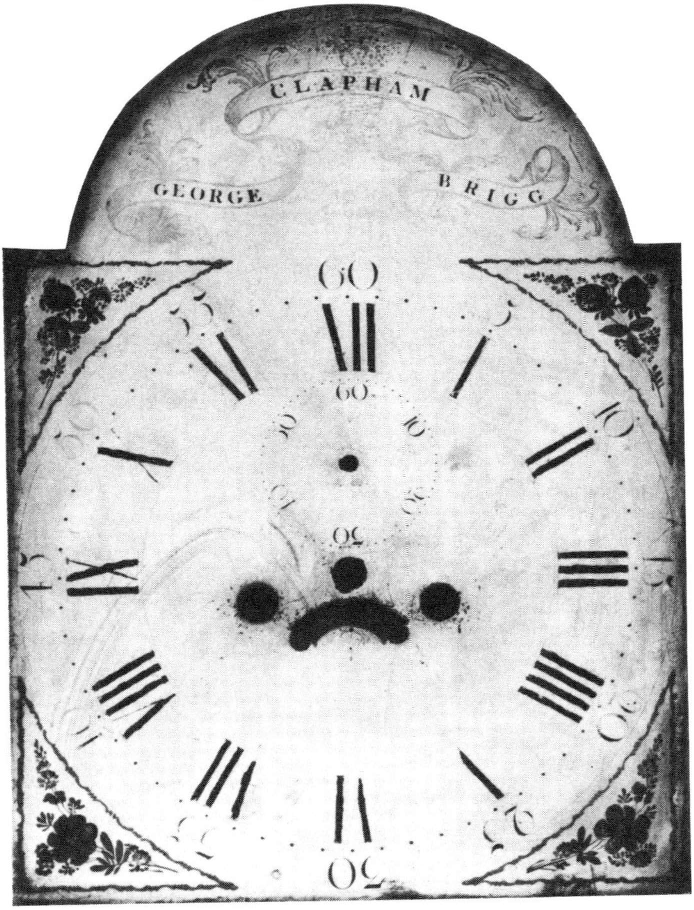

162 Furniture polish on dial edges have damaged this fine Wilson dial of 1785. See pages 233 and 235

over in ink. Any date or seconds rings should be done at this time, as well as any moons age semi-circles. Then the dial is put back and the turntable and the straight lines of the Roman numerals done with the aligning pen. At this point the dial can be considered as 'marked out'.

Maps can cause a lot of trouble because it is not generally realized just how many types there were. Take, for example, Wilson maps. The very earliest **(Illustration No. 22)**, is of the cut-off variety, with latitude and longitude lines very close together and not much detail. The ecliptic on the Americas is to the north of the equator. The second type is a bit later **(Illustration No. 23)** and this map has the lines more widely spaced as well as including more detail, but the ecliptic on the Americas remains to the north. Still later comes the slightly wider North America

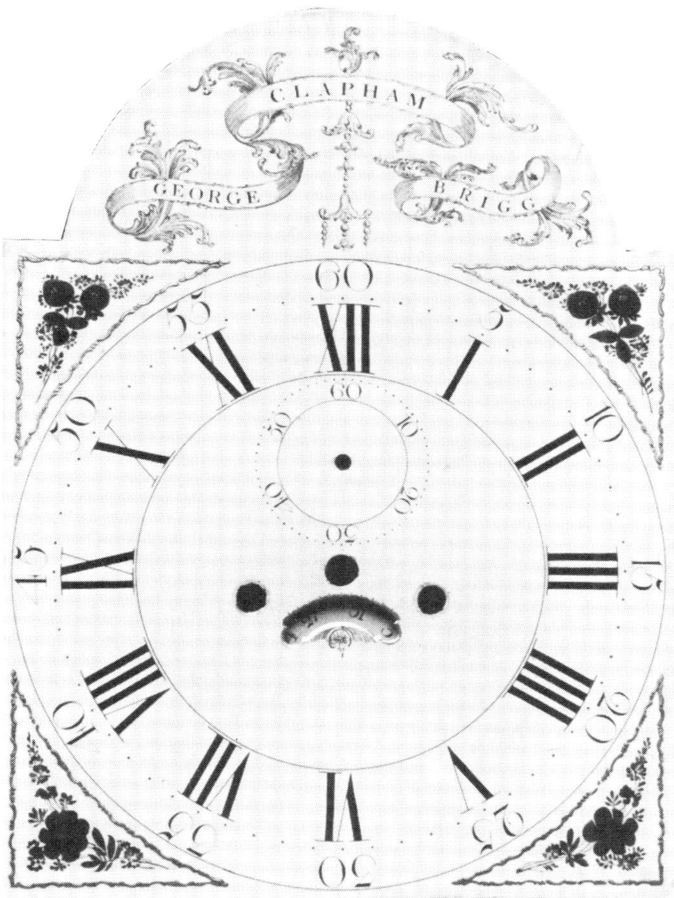

163 Cleaning has revealed more of the transfer printing in the break arch. The name George Clapham of Brigg was not transfer painted. See pages 233 and 235

(Illustration No. 24) where the ecliptic has changed to south of the equator. Of course there had to be different sizes of plates for these maps according to the size of the dial, but it seems essential when making a new dial to match the maps to the style and period of the dial, i.e. not to put late Finnemore maps on a Wilson style dial. Of course this degree of perfection cannot be attained without a great number of tracings. Photographs can be used where one has not got a tracing of a particular size, and one enterprising firm, Meadows and Passmore, has produced a rub-on transfer of two late Finnemore maps, similar to those found on p. 218 of Brian Loomes's *White Dial Clocks*. No opportunity should be missed to trace maps in good condition, though doing this on a map of the complexity used by the Nicholas brothers (**Illustration No. 40**), can take some time. The maps can be traced onto the dial in the same way as the main tracing, using the 6B pencil, but they are best done with separate pieces of tracing paper. This procedure is for new dials only, or complete repaints. Where maps have been on a very scrubbed dial it is rare to find no hint at all of even the equator, but this can happen, so then one has to find a map of the right dial maker's date and size. The map work is usually done after marking out.

Gold leaf work such as around the date aperture can be started next, but if there is gesso work such as elaborate corners or break arch motifs this obviously comes first. Laying down gesso is not easy, and it is best practised quite a lot on a flat metal surface which has been gloss painted. A modern gesso mixture found to be satisfactory is: 1 level measuring teaspoon of plain powdered gelatin, dissolved in 5 fluid ounces of very hot water, stirred well until it is completely dissolved. When the mixture is lukewarm, a small amount, say ½ teaspoon, should be added to about 1 tablespoon of gesso plaster in an eggcup, and stirred vigorously with a glass rod until well mixed. The consistency must be so that it can be put on with a small brush but still hold its shape. It may take several sessions to do all four corners of a complicated pattern. The work must dry naturally for at least twelve hours before it is carefully varnished with a clear gloss varnish. It is not easy to do this work accurately, and it cannot be hurried or interrupted. The same mixture is used for 'framing' scenes such as the four seasons, but obviously these scenes must be completely painted first and allowed to dry before putting on the motifs. Gilding the gesso is best done using Japan gold size and transfer gold leaf. Once the work is really dry, the gold can be varnished, as this not only protects it but dulls the brilliance down a bit. Gold leaf has a habit of getting into the wrong places, so especially with new work great care must be taken. It is obviously best to keep the leaf in a box with a lid, and it must not get damp.

If there is a lot of decorative artwork this is usually done next as it can take some time to dry, and it must not be varnished until it *is* dry. After this comes the painting of the Arabic numerals, including any date or seconds numerals, and the name plus town. The last thing to be done is the filling in of the Roman numerals. One type of dial made by Walker and Hughes had brass filings set in adhesive in the corners (**Illustration No. 51**). If a new dial of this sort is to be attempted quite a lot of practice is needed, as the technique is not easy. Not only the size of the filings is important, but also the colour of the brass. Silver leaf is troublesome to work with, as it is much thicker than gold leaf and tarnishes easily. Wherever possible, it is better to use gold leaf. Tin leaf might be a better substitute for the silver, but so far this has not been tried out. The final touch is either replacing the old cleaned up grommets or providing new ones, unless the dial never had any of these at all. Of course some really early dials had gilded winding holes. An inscription should be painted on the back of the dial with the restorer's name, address and date.

IDEAL BACKGROUND FOR DIAL RESTORER OR STUDENT

At this point it might be useful to know what sort of background a dial restorer/conservator should have. A thorough grounding in fine art with particular attention to oil painting as well as graphics is necessary. Cartography is very useful, as is picture restoring. It is no profession for those artists with a pronounced style, as one has to be a good copyist, imitating various styles.

STORAGE OF PAINTED DIALS

Old unrestored dials need equable temperatures, and should not be positioned where they can become even more dried out. Restored dials should, wherever possible, be on their respective movements and placed where there is good natural light, and not in a dark cupboard. This is because some oil paints discolour in the dark. They must not ever be dusted, only puffed at with a puffer brush. Spray furniture polish or spray window cleaner should never be used on the case, nor should the brass grommets be polished. Thermal shock must be avoided as well as direct sunlight, which rapidly fades some of the more fugitive colours such as crimson lake.

This book has taken two years to write and sixteen years to prepare. I have worked alone with the help of my husband in dealing with mechanical problems. He has also done all the photography and printing. It would be impossible as well as inadvisable to list each individual client and all the firms and other bodies that have provided me with a seemingly endless supply of dials for restoration, but without them this book could never have been written. Some people may be surprised to see 'their' clock illustrated, but others may be disappointed not to be included. However, I often have had to pick out the best photograph showing certain features. There are bound to be some inaccuracies or arguable points in this book but, to the best of my knowledge, this volume contains a true picture of a much neglected form of a unique art.

Bibliography

Horology

Allbright, Frank P., *Johann Ludwig Eberhardt and his Salem Clocks* (University of North Carolina Press, Chapel Hill, Winston Salem 1978)
Avery, Amos G., 'New England Clocks at Old Sturbridge Village' (The J. Cheney Wells Collection), 2nd edn (Old Sturbridge, 1966)
Bailey, Chris H., *Two Hundred Years of American Clocks and Watches* (Prentice-Hall, Englewood Cliffs, New Jersey, 1975)
Baillie, G.H., *Watchmakers and Clockmakers of the World*, 3rd edn (NAG Press, London, 1974)
Bates, Keith, *The Clockmakers of Northumberland and Durham* (Pendulum Publications, Morpeth, Northumberland, 1980)
Battison, Edwin A. and Kane, Patricia E., *The American Clock 1725–1865: The Mabel Brady Garvan & Other Collections at Yale University* (New York Graphic Society, New York, 1974)
Britten, F.J., *The Watch and Clock Makers Handbook: Dictionary and Guide*, 11th edn (Baron Publishing, Woodbridge, Suffolk, 1976)
Brown, Colin & Mary, *The Clockmakers of Llanrwst: Pre-Industrial Clockmaking in a Welsh Market Town* (Bridge Books, Wrexham, Clwyd, 1993)
Bruton, Eric, *Antique Clock and Clock Collecting* (Hamlyn Publishing Group, London, New York, Sydney, Toronto, 1974)
—*The History of Clocks and Watches* (Orbis, London, 1979 & Crescent Books, New York, 1982)
—*The Longcase Clock*, 2nd edn (Hart-Davis, MacGibbon, St Albans, Hertfordshire, 1976)
Cescinsky, Herbert & Webster, Malcolm R., *English Domestic Clocks* (Chancery House, Woodbridge, Suffolk, 1976)
Edwardes, Ernest L., *The Grandfather Clock*, 3rd edn (John Sherratt, Altrincham, Cheshire, 1976)
Elliott, Douglas J., *Shropshire Clock and Watchmakers* (Phillimore, London & Chichester, 1979)
Fennell, Geraldine, 'A List of Irish Watch and Clockmakers' (Stationery Office, Dublin, 1963)
Harlow, Samuel Boulton (compiled by Charles Aked & Rita Shenton) 'The Clockmaker's Guide' (Chronos, London, 1978)

Hewitt, P.A., 'The Deacon Family of Leicestershire Clockmakers' *Antiquarian Horology* (September 1986, December 1986 & June 1987)
—'The Deacon Clockmaker's Records' *Antiquarian Horology*, September, December 1987)
Hoopes, Penrose R., *Connecticut Clockmakers of the Eighteenth Century* (The New England Publishing Company, Stratford Connecticut, 1975)
Hudson, Felix, 'Scottish Clockmakers: A Brief History up to 1900' (Felix Hudson, Dunfermline, 1984)
—'Scottish Longcase Clocks 1780–1870', *Antiquarian Horology*, monograph no. 22 (Winter 1976–Autumn 1980)
Hudson, Felix & Norgate, Jean & Martin, *Dunfermline Clockmakers up to 1900* (Felix Hudson, Dunfermline, 1982)
Jagger, Cedric, *Royal Clocks: The British Monarchy and its Timekeepers 1300–1900* (Robert Hale, London, 1983)
Lloyd, H. Alan, *The Collectors' Dictionary of Clocks* (A.S. Barnes, South Brunswick, New York, 1964)
—*Old Clocks*, 4th edn (Ernest Benn & Dover Publications, London & New York, 1970)
Loomes, Brian, *Complete British Clocks* (David & Charles, Newton Abbot, Devon, London & North Pomfret, Vermont, USA, 1978)
—*Grandfather Clocks and Their Cases* (David & Charles, Newton Abbot, Devon, 1985)
—*Watchmakers and Clockmakers of the World, Vol. 2*, 2nd edn (NAG Press, Colchester, Essex, 1989)
—*The White Dial Clock* (David & Charles, Newton Abbot, Devon, 1977)
—*White Dial Clocks: The Complete Guide* (David & Charles, Newton Abbot, Devon, London & North Pomfret, Vermont, USA, 1981)
McKenna, Joseph, *Watch, Clock and Dialmakers of Birmingham, 1547–1900* (Pendulum Press, Hall Green, Birmingham, 1988)
Mather, Harold H., *Clock and Watch Makers of Nottinghamshire* (Friends of Nottingham Museum, 1979)
Moore, C.N., 'Chester Clocks and Clockmakers' (Grosvenor Museum, Chester, 1979)
Osborne, Clive (compiled and edited by James Heaps), 'Essex Clock and Watchmakers' (printed at Billericay, Essex)
Palmer, Brooks, *The Book of American Clocks* (Macmillan 1950)
—*A Treasury of American Clocks* (Macmillan Publishing & Collier Macmillan, New York & London, 1967)
Pearson, Michael, *The Beauty of Clocks* (Crescent Books, New York, 1978)
Peate, Iorwerth C., *Clock and Watchmakers in Wales*, 3rd revised edn (Welsh Folk Museum, National Museum of Wales, St Fagans, Cardiff, 1975)
Potter, Beatrix, *The Tailor of Gloucester* (Frederick Warne, 1903)
Pryce, W.T.R. & Davies, T. Alun, 'Samuel Roberts, Clock Maker: An Eighteenth-Century Craftsman in a Welsh Rural Community' (Welsh Folk Museum, National Museum of Wales, St Fagans, Cardiff, 1985)
Robinson, Tom, *The Longcase Clock* (The Antique Collectors' Club, Woodbridge, Suffolk, 1981)
Rose, Ronald E., *English Dial Clocks* (The Antique Collectors' Club,

Woodbridge, Suffolk, 1978)
Seaby, Wilfred A., 'James Wilson, Clockmaker of Belfast' *Antiquarian Magazine* (June, 1983)
—'Clockmakers of Warwick and Leamington to 1850' (Warwickshire Museum, Warwick, 1981)
Smith, Alan (Ed.), *The Country Life International Dictionary of Clocks* (Country Life Books, 1979)
—'Clocks and Watches: The Connoisseur Illustrated Guides' *Connoisseur* (1977)
—*The Guinness Book of Clocks* (Guinness Superlatives, Enfield, Middlesex, 1988)
Smith, John, *Old Scottish Clockmakers from 1453–1850*, 2nd edn (E.P. Publishing & British Book Centre, New York, 1979)
Stuart, Susan E., 'First White Dial Longcase Clocks and Their Cases', *Antiquarian Horology* (December, 1982)
Tennant, M.F., 'The Painted Dial in Ireland', *Antiquarian Horology* (Summer, 1991)
Theelke, Anthony, 'Faces of Mystery: English Decorated Enamel Clock Dials of the Eighteenth Century' (Anthony Theelke, Bristol, 1983)
Treherne, A.A., 'Nantwich Clockmakers' (Nantwich Museum, Cheshire, 1985)
Treherne, Alan, 'The Massey Family: Clock, Watch, Chronometer' (Borough Museum, Newcastle-under-Lyme, 1977)
Tyler, E.J., *American Clocks for the Collector* (Robert Hale, London, 1981)
—'Clockmakers of Sussex' (Publishers' Watch & Clock Book Society, Ashford, Kent)
Walker, John E.S., 'Hull and East Riding Clocks and Watches and Their London Origins including a Directory of Their Makes pre 1900' (Hornsea Museum Publications, 1982)
Wood, Stacy B.C., Jr. & Kramer, Stephen E., III, *Clockmakers of Lancaster County and Their Clocks 1750–1850)* (with a study of Lancaster County clock cases by John J. Snyder, Jr.) (Van Nostrand Reinhold Company, New York, 1977)
—'The Hoff Family: Master Clockmakers of Lancaster Borough' (The Lancaster County Historical Society, Lancaster, 1977)

Fine Art Restoration and Conservation

Bromnelle, Norman, Smith & Perry (eds), *Conservation and Restoration of Pictorial Art* (Butterworth & International Institute for Conservation of Historic & Artistic Books, 1978)
Keck, Caroline, *A Handbook on the Care of Paintings* (The American Association for State and Local History & Watson Guptill Publications, 1965)
Mayer, Ralph, *The Artist's Handbook of Materials and Techniques*, 3rd edn (London, 1973)
'Murray Pease Report: Studies in Conservation', Vol. 9, No. 4 (International Institute of Conservation, August 1964)

Ruhemann, Helmut, *The Cleaning of Paintings* (Praeger, New York, 1968)

Costume, Natural History, Art

Abbott, R. Tucker, *Seashells of the World* (The Golden Press, Western Publishing Co., New York, 1962)
Bossert, H.T. & Zschietzschmann, W., *Hellas and Rome: The Civilisation of Classical Antiquity* (Ernst Wasmuth, Berlin, 1936)
Byrde, Penelope, *Nineteenth Century Fashion* (B.T. Batsford, London, 1992)
Corbet, G.B. & Arnold, E.N., *A Handguide to the Wild Animals of Great Britain and Europe*, 2nd edn (Treasure Press, London, 1985)
Cunnington, Phillis & Lucas, Catherine, *Occupational Costume in England from the Eleventh Century to 1914* (A. & C. Black, 1976)
Gibbs-Smith, Charles H. & Victoria & Albert Museum, *The Fashionable Lady in the Nineteenth Century* (HMSO, 1960)
McKerrow, Mary, *The Faeds: A Biography* (Cannongate, 1982)
Nunn, Joan, *Fashion in Costume 1200–1980* (The Herbert Press, 1984)

Appendix

Suppliers

Chemicals, solvents, resins and lacquers

Archival Aids Ltd, P.O. Box 5, Spondon, Derby DE2 7BP

Gold transfer leaf: yellow gold, lemon gold & white gold

W. Habberley Meadows Ltd, 5 Saxon Way, Chelmsley Wood, Birmingham B37 5AY

General manufacturers of artists' materials

Daler-Romney Ltd, P.O. Box 10, Bracknell, Berkshire RG12 4ST
Windsor & Newton, Colart, Whitefriars Avenue, Harrow, Middlesex HA3 5RH

Dry transfer maps for hemispheres (late Finnemore pattern)

Meadows & Passmore, Medmaw House, Farningham Road, Crowborough, East Sussex TN6 2JP

UV lamp available from most electrical wholesalers

Index

Numbers in italics indicate illustrations